PARIS WEST TO PARIS

PARIS WEST TO PARIS

Around the World in Three Years with a Stop for War

ALSTON M. BARROW

Cimbarr Press
Tampa, Florida

Paris West to Paris:
Around the World in Three Years with a Stop for War
© 2024, Alston M. Barrow. All rights reserved.

Published by Cimbarr Press, Tampa, Florida

ISBN 979-8-9915167-2-3 (hardcover)
ISBN 979-8-9915167-1-6 (paperback)
ISBN 979-8-9915167-0-9 (eBook)
Library of Congress Control Number: 2024920978

Publication managed by AuthorImprints.com

To my forgotten kinfolk

CONTENTS

PREFACE

I HAVE OFTEN WISHED I knew more about my ancestors. Maybe I can inspire a future Barrow to set pen to paper. Twenty-first century life is flooded with books and data, so it's easy to worry about adding to the mountains of recorded stuff. But people have always been served by the permanence of books. I have notebook records and some financial newsletter writing background, so I offer here my travel experiences. I have altered most of the names to protect both sides. Diary notebook entries have been italicized.

Please keep in mind this is a memoir of my three years traveling the world in the 1960s, including my time in Vietnam, a small piece of my life—less than 5 percent—before I was even married. I am eighty-four-years-old as I write. In an Afterward, I offer details of my upbringing—especially for my eight grandchildren here in Tampa, as I have never been much of a storyteller. This memoir has four parts: my travels including a year in Vietnam, my background of Navy service, some tales about my youth, and reflections on post-bellum life by a Barrow relative.

Except for Martha Barrow Turnbull's rambling thoughts on gardening and recipes, I hadn't run across many Barrow ancestors who bothered to document anything since Bennett Barrow's 1838 diary, published by Columbia University Press in the 1940s. I must acknowledge the huge contribution of William Barrow Floyd's genealogical research in writing *The Barrow Family of Old Louisiana,* with 500 copies printed in 1963. I was impressed by two points he discussed in detail. The first is his commentary about the loss of so many family records and furnishings by fires and by moves from place to place with displacement of family members over time. The second is the inevitable loss of accuracy when family history and traditions are verbally passed on from one generation to the next.

Upon discovering that I was in the process of making my contribution to the Barrow family papers, family members came forth with old photos and memorabilia. Two surprises were *My Autobiography to My Sons*, a brief work by my paternal grandfather Ruffin Bennett Barrow, and a longer story, in effect a personal memoir, by his sister, my great-aunt Hazel Barrow.

Most of our North American ancestors migrated west to Louisiana from families starting in Virginia, moving south to central and coastal North Carolina, and then westward. There's an old Barrow home on the James River. The Barrows acquired Spanish land grants in the Feliciana parishes around Saint Francisville and expanded land holdings across the state and in Tennessee. There have been Barrows who were explorers, westward pioneers, millionaires, rebellion fomenters, confidants of American presidents, and military leaders. We've even had a US senator! I wish we had more firsthand data rather than outside records and verbal recollections destined to fade away over time. Lots of untold stories.

First, I'll dispose of that famous celluloid mate of Bonnie, the bank pillager Texan Clyde Barrow. He's not even a distant cousin. But I'm related to William Barrow, who was among the small group of men who founded the Republic of West Florida, capital at Saint Francisville, where the Bonnie Blue flag flew as the Republic's official symbol of independence (albeit short-lived) from Spain.

That 1810 flag was the first lone star white flag on a blue background, followed nearly thirty years later by a similar version created for the Lone Star State. The citizenry of Texas, perhaps a bit conflicted and defensive because they were a nation before becoming a state, would likely take umbrage at my telling the story of the first lone star flag. If you drove from Saint Francisville west across Texas telling the story of the Bonnie Blue flag, all the way across its rattlesnake-infested Panhandle to the safety of the New Mexico or Oklahoma state line, you might risk assault or even assassination by some big-hat no-cow cowpoke or maybe a sideswiping by a cowgirl in her Ram Charger. Incidentally, Alaskans like to say, if you cut their state in half, Texas would be the third largest state.

My father's brother and other forbears made Texas their home, including R. R. Barrow who owned several plantations there. Senator Alexander Barrow made a windy four-page speech delivered to the United States Senate on the admission of Texas to the Union in 1845. Maybe being located next door to fun, food, fishing, dancing, and Mardi Gras has a unique impact on Texas.

When I was in the military, I started carrying a pocket diary, a British Smythson of Bond Street three-by-five-inch featherweight onionskin notebook. I made notes in them almost daily. The early ones I put aside and never looked at again. After returning to the real world—work, marriage, family, and such—I continued carrying a pocket diary all the way to arrival of Steve Jobs's Apple, a revolution in life's ways. My old diaries gathered dust. I didn't want to revisit the past.

Fast forward half a century, and I took a bike tour through Vietnam. Coming home, I decided that if I'd taken the trouble to make all those years of notes, for whatever reason—misguided or not—then I owed it to myself or perhaps curious relatives to read what's there. So I did. As the old song goes, I forgot to remember to forget. I found an enormous amount of detail—much mundane, some inscrutable or forgotten, but a lot that triggered immediate recall. Curious, perplexing, and some fascinating, memories long shelved were instantly revived. What possesses people to jot daily details down? Forgetfulness, force of habit, whatever works, but keeping a diary is a traditional pastime.

```
AMEXCO A PARIS
161 0148
BLOKOUS A PARIS
CXA957VV    FGA347 EDC154CED666
RR RUFGWC
DE RUCKC 423 09/1906Z
ZNR
R 091750Z
FM USS GEORGETOWN
TO RUFGWC/LTJG ALSTON M BARROW C/O AMERICAN EXPRESS CO 11 RUE
SCRIBE PARIS FRANCE
INFO ZEN/BUPERS
US GOVT NAVY GR59
BT
UNCLAS
ORDERS RECEIVED FOR YOUR DETACHMENT. YOU AR CSCHEDULED TO
REPORT SAN DIEGO, CALIF., NOT LATER THAN 6 JULY FOR ONE
WEEK SPECIAL COURSE WHILE E NROUTE NAVAL ADVISORY STAFF
SAIGON. RECOMMEND RETURN GEORGETOWN NOT LATER THAN 25 JUNE
I NCIDENT RELIEF AS FIRST LT. FOR COMMAND AND BUPERS PLAN-
NINO REQUEST ACKNOWLEDGE RECEIPT THIS MESSAGE EARLIEST.
FURTHER, PROVIDE INTENTIONS WHEN FIRM.
BT


NNNN      
AMEXCO A PARIS
BLOKOUS A PARIS
```

Orders for Detachment telegram

VIETNAM

ONE DAY IN MAY 1965, when I was twenty-four-years-old, the captain of my ship passed around a bulletin from the Pentagon asking for volunteer officers for duty as military advisers in South Vietnam. With my experience as a gunnery and deck officer I figured that I was a prime candidate. I was ready for bigger adventure than duty on an independent steaming spy ship. The situation in South Vietnam was fast deteriorating and I sensed that the buildup of American forces under a bulldog president like Lyndon Baines Johnson (LBJ) was inevitable. I volunteered.

We were back in Norfolk for a break after a long independent steaming deployment to South America. The captain called me up to his stateroom and said, "You're the only fish that bit." He asked me about my transfer request. I told him that I had really enjoyed my time under his command on the USS *Georgetown* (AGTR-2). The months doing surveillance work on both coasts of South America and in Cuba had been exciting—it was way better duty than my first ship—but I would like to get nearer a war.

He told me, "Well, you're a Louisiana guy. You hunt and fish, whatever. You've been a deck and gunnery officer and had navigation experience and you're—I won't say crazy but adventurous enough—that I won't be surprised at all if you get selected for advisory duty in Vietnam. You probably want a few days at home since you might not see it for a while." I thanked him and repeated how much I enjoyed serving under his command. I started thinking that things may be changing fast. Like getting exciting. If I went to war, I might not come home. I might never go to the places I planned to see, do the things I dreamed of doing.

I had a general idea of what being an "adviser" entailed based on media reports filtering into Saigon bureaus from advisory teams in the field. The term "team" was a loose approximation of an idealistic objective. *Our Navy* magazine noted, "Junk force advisors are probably the most isolated sailors in Vietnam. Some 100 advisers serve with twenty Vietnamese Navy costal groups located along 1,000 miles of coastline in South Vietnam. Many of these bases are accessible only by air or water."

As I soon discovered, many of those bases never reached or maintained their full four-man adviser team strength. Frequently, advisers patrolled without their counterpart, the Vietnamese officer of similar rank serving alongside the American adviser. What's more, going on patrol generally entailed using scrounged-up equipment and hand-me-down surplus arms. It was a strange conflict where you rarely knew who was friend or foe—people you served with might be playing both sides.

The buildup to open warfare had begun in South Vietnam, and I volunteered for duty there. There were no dreams of heroism, honor, or duty. That was movie stuff. I was anticipating a great life adventure worth the cost of some personal risk—life is full of risks.

READING, ROLLING, AND WATCHES

OUR DEPLOYMENTS TO SOUTH America had been days and days surrounded by water, independent steaming, and often empty horizons. The long stretches of time off watch and between operations and training exercises were filled with reading, coffee, scuttlebutt, cards, and more reading, Reading was a big pastime at sea, a solitary alternative to playing rounds of bridge and acey-deucey, watching wardroom Westerns (a.k.a. shit-kickers) or the occasional cinema classic, or some personal version of contemplating one's navel—grousing, scanning the horizon, writing home, scratching itches. Even nineteen-year-old dolts who had never faced a pulp paperback become addicted to stories in print after a few months of sea duty with no bars, booze, women, cars, or bowling balls.

I had pretty much absorbed most of the works of Lost Generation Paris notables like Ernest Hemingway (Hem), T. S. Eliot, Pablo Picasso, and F. Scott Fitzgerald. I had taken French at Vanderbilt. It was obvious: go to Paris on my leave instead of heading home to St. Francisville to hang around Spanish moss and Civil War detritus. Romantic notion—I needed to see Paris before I was dead. The action was in Paris! I must wander the streets of the Left Bank where Hem lived before my time was up.

Paris! Where the most interesting people in the world played and partied, where Hemingway lived with his first two wives.

I was only nine when William Faulkner won the Nobel Prize for Literature; Hemingway won when I was in high school. Publicity followed—*Life* magazine and newspaper profiles, even an *Old Man and the Sea* movie. They coexisted; I went through school reading works from both.

My fourth and final year of duty as an adviser would be under Military Assistance Command Vietnam (MACV). That's about all I knew. Doing what and where were big unknowns. With forty-six days accumulated leave from being at sea in South America and the Caribbean, I put in for two weeks leave and bought a round-trip ticket to Paris.

I don't remember the flight to Paris, but it must have been on a commercial airline, not a military hop. I was *enthralled* by actually being in Paris. I wandered the streets, the Left Bank, the Champs-Élysées, shops, parks—the Seine was just as I had imagined. Some young Germans on vacation were also wandering and we linked up to continue exploring the city.

A week or so into my leave, I picked up a Western Union wire from my ship at the American Express office notifying me that I had been assigned as an adviser with MACV—Vietnam. Poignant moment! I walked out of American Express and wandered dazed down the Rue Scribe pondering. My life was fast changing. I was to report to Oakland (CA) Naval Supply Depot for temporary duty. The pre-deployment standard courses were two weeks of survival school—eat snakes, start fires, and such. This sounded intriguing with my Louisiana country roots. Since I would be operating with the Vietnamese, a crash course in the Vietnamese language would be added. The present intruded: June 1, I made a note to write "German girl" whom I had met in Paris.

On June 11, the day before my birthday, the officers threw me a farewell party. I recall being surprised and taken aback by my commanding officer's courtesies and civility—these were not the usual military ways. The custom aboard Navy ships too often was a callous here-today-gone-tomorrow feeling. The *Georgetown* was a squared-away ship and a great duty station. Still, I was eager to head out for my great adventure.

My new diary was full of notes on preparations for leave to Paris, followed by preparations for Vietnam. The Naval Advisory Group was undermanned, and the dispersed junk bases on the coast were vulnerable to attack. I was needed pronto because the junk bases were shorthanded. The weeks of temporary duty for survival school training and Vietnamese language classes were canceled. Schedules all down the line were accelerated or scratched in the scramble of a rapid buildup after the Gulf of Tonkin incident.

In 1964, two Navy destroyers claimed they had been fired on by North Vietnamese forces. In response, LBJ obtained permission to increase US military support of South Vietnam. Years later, declassified documents revealed that LBJ was faced with an eight-fold buildup of men, arms, and material from the North in the twelve months after President John F. Kennedy's assassination. The invasion was on. It came along the Ho Chi Minh trail and via the South China Sea as verified from top secret across-border monitoring. A base had been mortared and overrun and the senior naval adviser killed. I was not great at learning languages, but I soon realized how helpful a little more fluency in Vietnamese would have been out on patrol. Only my counterpart officer could speak fluent English.

I managed to squeeze in my week's US farewell wandering through San Francisco. The first day there, I reported for temporary duty at the Oakland Supply Depot. The warrant officer on duty, knowing I was headed for combat, took pity on me and told me not to bother checking in each morning. I could go have my last fling closing a few bars. One afternoon I went down to the Abercrombie & Fitch store—back then a hunting and safari outfitting store—and bought a little Browning .25-caliber automatic pistol and the 1966 edition of their onionskin diary for $2.75. I already carried the 1965 edition everywhere.

On one of the first diary pages in 1965, I calculated the boost in pay awaiting me for serving in a combat zone. As a 0–2 grade officer my monthly base pay was $462. Now they would tack on $55 monthly combat pay, $51 cost-of-living reimbursement (why?), $43 subsistence supplement (don't ask what for), and $40 imputed value of my income tax-free status during my year (or however long I would be in combat). I have always been a numbers and statistics guy; it's an ingrained habit.

I spent one year and ten days in Vietnam, save for the one rest and recreation (R&R) leave break in Hong Kong—that was two days travel and five days in Hong Kong. My assignment was serving as an adviser in the field at Vietnam Navy junk bases patrolling in the South China Sea in II Corps, the centrally located corps of the four geographic divisions of the country, from the North Vietnam south to the Mekong Delta. After returning from a bike trip to Vietnam in 2012 with Vermont Bike Tours, that's when I wiped the dust off my half-century-old notebooks and began reading.

EARLY JUNK PATROLS

JULY 26, 1965. *Departed US on Continental flight.*

The Navy flew me west on Continental Air via Subic Bay to Saigon. As we approached the coast over the South China Sea, the speakers played the sentimental World War II ballad "I'll Be Seeing You." The GI next to me was moved to tears: Going to war, bye Mom! I reported to Naval Advisory Group, MACV Saigon on August 2, 1965, and was issued black pajamas, a black beret sporting a tin Yabuta (Vietnamese boat) junk pin, two sets of Army fatigues and jungle boots, a .45 pistol and a .35 Carbine. I took a C-47 up to Nha Trang on the coast in II Corps and from there I went about thirty miles by junk to my base on Binh Ba island just south of Cam Ranh Bay.

> **AUGUST 14.** *Saigon. Note to buy from Nha Trang PX* [post exchange]: *flashlight, sandals, toilet paper, wide brim hat, black dye, candles* [things I needed in the field and couldn't count on the Navy to provide any longer].

> **AUGUST 17.** *Arrived Binh Ba, my junk division's base. BITCHES: out of paper; no training ammo; request L-19 flights twice weekly for CO* [commanding officer]; *fuel oil, kerosene not received for a month.*

Binh Ba Island, old French Foreign Legion
outpost, my first junk base

I no longer recall what some of my entries mean but others are firmly cast in my memory. Daily notes in my little onionskin diary during the initial weeks at my new base were skipped as this was a whole new ballgame for me. All my focus was on new people, new duties, new weapons, new food, new patrols, and making sure to keep my body parts working. After I ended up in the Army field hospital tent New Year's Eve, I made more time for putting my diary back in action and resumed my regular note-taking.

MY ISLAND JUNK BASE

MY FIRST DIARY ENTRIES at the base were radio frequencies for communication and monitoring US ships patrolling offshore in the South China Sea. We monitored activity—trying to interdict Vietcong (VC) supplies moving by sea—we hoped that a ship might come to our assistance if we needed help. BEE KING was the call sign of our destroyer on patrol in II Corps. I had my list of stuff to buy whenever I got up to Nha Trang. Days later, my wish list expanded to swim equipment (the reefs were teeming with fish and spiny lobsters), tennis shoes, French books, green T-shirts, and bigger black pajamas (the prescribed night patrol uniforms at sea).

The base had about fifty Vietnamese sailors, two junior officers, and perhaps a dozen wooden diesel junks. The junks, Yabutas, are listed at thirty-six feet and carrying one .30 caliber plus one .50 caliber machine gun- we had neither. I recall our base had little more than the crew's weapons, old Army of the Republic of (South) Vietnam (ARVN) open-sight semi-automatic rifles. I don't even remember seeing a command junk or a South Vietnamese Navy (VNN) sail junk in II Corps. According to varied records, there were 389 motor junks and ninety-five sail junks in mid-1965, 488 junks in 1966, and only 290 junks by 1967 patrolling the 1,200 mile coastline.

The Yabutas weren't much of a combat vessel, but were superior to the competition manned by VC out on the South China Sea. They sported a three-cylinder Gray Marine diesel engine and an open-windowed wood centerline cabin to protect life vests, plus tools, a hibachi pot, food, water, and ammunition (ammo) from the seas and weather. The open cockpit was at the back of the cabin. We had a U-shaped stack of sandbags in the bow for armor. Later when we got tripod-mounted Browning machine guns, they were placed inside the stacked sandbags.

We hoped that the island location for the base gave us more protection from VC attacks than most junk bases. Our only power was a small diesel generator for running our radio (call signs were Baron60 for forward air control (FAC), Bluefin for medical evacuation (medivac) chopper, and for contact with the communications (comm) center in Nha Trang. We had just enough extra power for two or three light bulbs in the evening. We scrounged up a reel of concertina wire to barricade most extra window openings. The Kentucky-style one-hole shitter was twenty-five meters out back, halfway up the hillside. I remember sitting out there on full moon nights, AR-15 on my lap, thinking of dumb ways to die and the inconvenience of exotic diets and dysentery.

Our sleeping quarters were at one end of the second floor of the long-abandoned brick French Foreign Legion barracks. The Vietnamese sailors and my counterpart were housed in another barracks across the little harbor. There were stairs on each end of the building—we built stairway doors out of boards and concertina wire to barricade ourselves in at night and alert us to any surprise visitors. Those were the days before $5 motion-sensor lights.

We placed our sleeping cots randomly along the wall on each side of the room so we could cover each other by looking across to the open windows on the other side. Small comfort to be sleeping with the next guy fifteen yards away. After nightfall, we locked the stair gate and set up our loaded weapons on the deck at our bunks. Grenades were next to our rifles. We each had a large open window by our bunk, so we could fire out the window along the outside walkway. We could also jump through the window and spray enfilading fire inside the room if the enemy got inside.

We were stranded in the tropics without cold drinks or Navy chow. I started walking my counterpart, Mr. Tuong, to the fishing village at the other end of the island in the evenings. The village's hooch bar offered Ba Muoi Ba, Beer 33, with ice. We always had a table on one side of the hooch. Lieutenant (Lt.) Thieu, the base boss, warned us that the table at the other end was for VC drinkers. Since we walked single file on a dirt trail through trees after dark, we always went fully armed. After a few weeks of nervous beer round trips, we managed to get an old two-cubic foot French refrigerator for beer and food. It cycled refrigerant fitfully on an open flame kerosene wick at the back of the unit, lowering the temperature to about 45°F.

My counterpart base commander 2nd from left,
me in beret escort Navy commander checking out
our island for future Swift boat home base.

THE US ADVISORY TEAM

A FULL ADVISORY TEAM consisted of two junior officers and two senior enlisted, but the coastal surveillance advisory program was new; most bases were short their full four-man US advisory complement. I suspect that I had been rushed to Binh Ba after Navy brass on a photo-op tour of the base with Secretary of Defense Robert McNamara, General William Westmoreland, diplomat Henry Cabot Lodge Jr. and an assortment of admirals were questioned by the press corps about the personnel shortage. The senior naval advisor (SNA), an old mustang (officer commissioned from enlisted ranks) lieutenant nicknamed "King Bobby of Binh Ba" for his flamboyant ways, gave them the full tour dressed in our standard-issue black pajamas, beret, criss-crossed ammo bandoliers, carbine, and flip-flops.

On my first night patrol down the coast, we went into a long narrow bay known as VC-controlled territory. We had no business being there except to poke the hornet's nest. We took some small arms fire across the bow from the black shoreline and returned fire. At that time, I don't believe we even had Browning 30 machine guns for the patrol boats. King Bobby always wanted action and excitement. The trip was my introduction to Bobby-style coastal warfare. Bobby called up for aircraft fire support with no luck. Headquarters was probably on to Bobby's ways and we pulled out of the bay and headed home after no support showed up.

Looking back, I suspect Bobby was checking me out while having his fun and deciding whether he should try to finagle a new duty station. The buildup of Cam Ranh Bay under President Johnson was underway, with thousands of Army troops and tons of equipment pouring into the Bay. Our base sat on an island at the bay's entrance. We were shortly to be out of a job, no longer canaries in the coal mine guarding the bay from infiltrating VC junks.

MACV soon upgraded our name from lowly "junk divisions" to Coastal Patrol Groups as befit a Navy sporting nuclear subs, Mach 2 fighters, guided missiles, and the like. We also got more help in the way of more intense aerial surveillance and faster American-manned Swift boats along the coast. The enemy was having a harder time using the South China Sea.

The advisers I met during my year were: (1) looking for action, a promotion, a medal; (2) avoiding action, scared, resigned to fate; or (3) just adventurers at heart—carpe diem. King Bobby was type 1. Two later advisers I served with were type 2. I was type 3. A junk base usually had two junior VNN officers as counterparts to their American peers. King Bobby was not the first mustang officer I had worked with in my Navy years. Like horses, mustangs were feral animals and usually retained a wild streak after being civilized (broken). The early mustangs often were battlefield promotions like Audie Murphy, the WWII hero and movie star of numerous macho action films (shit-kickers in military slang). Bobby claimed to have been both in the Army and the Marine Corps—maybe bullshit, but I never found out as he was shortly to depart.

Bobby was a friendly, okay guy. I was soon flying patrols with the Army warrant officer twice weekly in L-19 spotter planes who was a different type of mustang. He had gotten his college degree as another route to commissioning and pay raises.

On the USS *Georgetown*, my immediate superior was an amiable old WWII vet mustang—except when in the company of the ship's communications officer, a loud and obnoxious jackass mustang who steered him awry with gossip and bullying of junior officers around the ship. Their view, partly correct, was that any officer coming straight from college or officer candidate school (OCS), with no prior military experience, was excess baggage in the hierarchy. They were extremely put out with annual training visits of cadets and midshipmen like me, where the "military" was often just another course like science and art history.

THE JUNK PATROL ROUTINE

KING BOBBY SOON NEGOTIATED a transfer to river patrol boats. After he left, we stowed the black pj's and boots in favor of fatigues, berets, and flip-flops. Our surveillance patrols gradually increased, usually running overnight two or three times a week. We returned to base at dawn, sleeping during the day. Sometimes we stayed out for two to three days. When the monsoon rains came, we stayed wet like bilge rats. My skin fungus spots became permanent (as I learned at the beach in Australia a year later).

Bobby's replacement soon reported to the base, a lieutenant from New England. The same month, I was also promoted to lieutenant, the equivalent of an Army captain. As a regular, presumably career, officer with a four-year obligation, I was subject to an automatic extension of duty for six more years unless I requested to resign my commission and decline naval reserve duty no later than six months before the end of my four years.

That time arrived. I sent in my letter of resignation intent on turning in my beret, carbine (useless and rusty), pistol, and ridiculously small black pajamas. I would take a physical, get mustered out (discharged) in Saigon, then fly out a civilian at my own expense. Not having the normal expense of flying my ass back to California, paying for my unused vacation leave, and such probably ensured that any senior officer would think twice before costing the Navy a hefty pile of mustering out expenses by turning down my resignation request.

I must have played my cards smarter than others. Decades later, I got a call in the middle of the night from the New England lieutenant who had tracked me down. He'd remained in the Navy, having extended his enlistment for six more years in a war that was to last with fits and starts until 1975. He was still just one rank higher up the ladder after reaching retirement. I didn't have the heart to ask him if his Navy career had gone to plan or had gone awry—the answer was obvious.

Back to our patrols. By the third day and night at sea in a rolling junk, I recall being shocked at myself for staying wedged against the gunwale in the bow snoozing as we pulled alongside junks to check for contraband: Let them shoot first, I'll just take my chances and keep using my rifle as a pillow! The closest we came to getting sunk was probably when we "attacked" a swimming deer and bent the prop. Back at the base, the fresh carcass was tossed on a bonfire—unskinned and ungutted. I woke up from my post-patrol nap to the stink of singed hair over half-raw meat instead of our first palatable red meat venison feast.

Loosey-goosey accidents are all too common in combat, and most don't get reported.

Killed in action (KIA) by friendly fire. Wounded in action (WIA) because the duty guard mistook a guy headed for the latrine for a VC sneaking in. Missing in action (MIA) because the patrol got lazy and walked right into an ambush. (Years later, I was taken aback by a trunk label on the first KIA car I saw, thinking, "Yikes, do those South Koreans realize how many new car sales they're going to lose to military families by sticking that name on their cars?") On one patrol, I had a grenade with a damaged pin and carelessly pulled the pin, tossing it just clear of the gunwale. Unluckily, it short-fused, rocking the boat. I pulled up the hatch cover and—thank God—no water rushed in.

I always wondered what would happen if I lobbed a grenade down the open hatch of a hostile fishing junk. If it was loaded with munitions instead of fish, it could have blown up and sunk us. I could only sling a grenade overhand about thirty-five yards. Ten sandbags in the bow weren't much protection from dumbass explosions. Even now, I still think, "Life can be short." You're tooling down the two-lane and a guy comes across the yellow stripe, you're either quick or dead, and you still need to miss the telephone pole by the ditch. When my grandchildren learn to drive, I tell them, "Now, as a driver, you must hereafter be eternally on guard and resolve to run off the road to avoid any chance of a head-on collision."

We lived or died, patrolled, and ate with the Vietnamese sailors, usually accompanied by my junior counterpart, a VNN junior officer. We would flag down fishing boats, pull alongside, and check for weapons and contraband. If the boat turned tail, we fired shots across the bow. My guess is that we managed to check maybe 20 percent of boats passing through our patrol area. Any coastal VC or sympathizers could have been alerting infiltrating boats of our comings and goings so we patrolled mostly at night with no lights, no smoking, no radar. We intercepted only small amounts of weapons or contraband, but even then we turned out to be pretty effective at reducing VC infiltration of material by water. By the end of 1965, MACV decided that the coastal surveillance program based primarily of junk divisions merited heavier support—US naval forces became directly involved.

At sea, almost invariably meals were boiled rice and squid in a hibachi pot on coals. On base, we ate soup, rice, squid, fish, and dog from the puppy pen on the base. With no screening, no fans, and no air conditioning, nasty black and green flies were everywhere—sharing and seasoning our soup and rice. When served, we would spend the first couple of minutes picking as many flies out of our food as possible. Where were the chickens and potbellied pigs? Ducks? I love Peking duck! We had home-made nuoc mam, a fish sauce of fish, salt, and peppers rotted for weeks in earthenware pots. One day at lunch, we received tubular pieces of boiled greenish translucent meat with trian-gular spine/ribs, which could only have been sea snake. Not tasty!

A few months into my tour, our chief gunners mate, a quiet beef-and-potatoes midwesterner who looked like actor Cary Grant, was pulled back to Saigon forty pounds lighter and from suffering with dysentery. For his farewell lunch, we toasted him with traditional raw chicken egg embryos in the shell. Chief tried to keep it down but puked.

Sometimes after sleeping off a night patrol or cleaning weap-ons, I would take my rifle and canteen in the midday quiet and work my way uphill behind the base, pushing through the brush and boulders. Like a deer, I stopped every minute or so, looking and listening in the shade of a boulder or tree, checking my back track. Reaching the top, I would find a view with a good rock to sit on. Ahead and far below the blue South China Sea stretched east toward California where I had come from, merging into a hazy sky. All was quiet. The only movement was a breezy leaf, a breaking wave far below, or the black speck of gulls sliding through the air out there. A beautiful and deadly sea shimmered on a silent clear day. The war, smoke, smells, and noise were gone for now, left far behind me on the other side of the island.

I went to 8th Field Hospital in Nha Trang with dengue fever on New Year's Eve.

THE NEW YEAR

DECEMBER 30, 1965. *I have dengue fever, Army 8th Field Hospital.*

JANUARY 21, 1966. *Tet holiday.*

JANUARY 21–24. *Buy fruit for Tet, service .50 cal mach gun, inspect bunkers, re-tile sentry posts, check fire .30s, emplace 60 mm.*

JANUARY 26. *Today over the hump.* [Since I departed San Francisco July 26, I must have calculated my one-year tour was halfway complete.] *Wetting-down party for my promotion. Drove injured Viet from Cau Da to field hospital.*

JANUARY 30. *Patrol tonite at 2000 alone.*

JANUARY 31. *Payday. Condition gray.*

I started jotting entries in my fresh diary for 1966 with the memoranda page listing fourteen radio call names starting with Bamboo Segment 33, the senior adviser for MACV, and the last one Bluefin, for medivac, a medical evacuation chopper. I remembered spending three days in an Army hospital tent in Saigon recovering from dengue fever. Upon reading my diary, I was shocked to learn that I had spent *eight* days in the field hospital tent up in Nha Trang on a cot in a tent with about fifty buddies.

On New Year's Eve, our line of cots shifted and some of us helped with Republic of Korea (ROK) soldiers landing on the tarmac in medical evacuation (medivac) choppers. There were "okay" guys, walking wounded, stretcher cases, and bodies. This was the first firefight of the war between ROKs and North Vietnamese regulars. A company or more of North Vietnamese regulars infiltrating south in the night collided with the ROKs. The ROKs, terrific fighters, chewed them up while taking quite a few casualties themselves.

As a Navy adviser independently attached to a South Vietnamese base, I was eventually discharged to go back to work by the Army nurse. I got a ride back to the port still feeling weird, and a junk from the base picked me up. It was the middle of the monsoon season and I resumed rainy night patrols. In hindsight, my recovery was surely incomplete at that point. I probably compromised my resistance to later infections, even though in the days ahead I gradually recovered my strength in spite of our crappy diet.

The base sailors carried bolt action ARVN rifles. Although the base had some protections as an island, we remained open to attack. In the fall of 1965, two or three bases were hit by mortar fire; one was sabotaged and overrun. I went to the Special Forces supply depot in Nha Trang, instructed by scuttlebutt to bring two-fifths of Chivas Regal from the PX, my monthly booze allotment. A master sergeant jumped on a forklift and brought out giant canvas hobo bundles onto the tarmac, dumping and opening them. The sergeant owned a string of bars in town through his mama-san (a woman in a position of authority; in this case, his girlfriend). He set the whiskey aside.

We picked out a couple of AR-15s (no traceable serial numbers), some .45-caliber grease guns, two .30-caliber machine guns, and an M-79 single-shot grenade launcher. For our barracks home and the base, we retrieved concertina wire and sandbags. The carbines we were issued had rusted quickly and were underwhelming, so we dumped them for the AR-15s, electric taping clips together with about ten rounds per clip, including one-third tracers (great secret South China Sea patrolling pleasure—test-firing clips with tracers into the night). In the military, field units constantly reinvent the wheel. By trial and error, we switched from thick Army gun grease to WD-40, and religiously broke weapons down and cleaned them after every patrol. I passed on a case of antipersonnel mines, having already seen so many disabled individuals, especially children, around the country.

LIFE GETS BETTER AT THE BASE

FEBRUARY 3, 1966. *Patrolled tonite off Hon Heo Peninsula—two suspects.*

FEBRUARY 5. *Full moon. No patrol. Finished reading Dr. Zhivago.*

FEBRUARY 12. *60-mm 140 he, fourteen ill., .50 cal. 14,400, .30 cal 120,000.*

FEBRUARY 17. *Firefight with VC junk sunk junk 7 KIA recovered no bodies*

Or material

FEBRUARY 21. *One VC body, will patrol area tonight.*

FEBRUARY 23. *Four more bodies, good patrol last night, Cdr. Toi kicked Thu's ass and he went out with me, detained some deserters and two VC suspects.*

When the Cam Ranh Bay Army base buildup started in the fall, a little Bell H-13 two-seat chopper dropped down to see the base. The pilot said he didn't have much to do because his Army general boss was afraid to fly in it. Our bay was bordered about twenty miles in every direction by hostile territory. He took me up a few thousand feet to 65°F above the base and we cooled off. He asked if he could get me anything, and I said how about a case of grenades and some beer. And that's how I got grenades for the base.

A few weeks later, he connected us with a sergeant in charge of warehousing supplies now pouring into the base by Caribou (military aircraft), truck, and ship. (Later, I counted seventeen Lykes cargo ships, a company I would know well in coming years, at anchor in the harbor). First available were surplus C-rations in cardboard boxes, equivalent to a one-step upgrade from Confederate salt pork and hardtack. The cute P-38 can opener and canned fruit cocktail were favorites. The crackers and lima beans were throwaways—even to us hard-up squid and soup types. Within weeks, we were picking up real canned food. Our big score was a whole case of Dinty Moore stew cans, which we polished off within a few days. I probably gained a pound or two of weight back on my svelte tropical frame.

As the war buildup progressed, we were able to get more mail and so-called care packages from home on a more regular basis by taking a junk up the coast to the US Navy Nha Trang compound. We received permission to stay overnight in the compound while the junkmen refueled and went home, or drinking and whoring, or whatever they did in their time off—God knows. My care packages were the best because Mother well understood my craving for an adequate stock of paperbacks. I took first choice but had to share. She also would stuff in snacks of all sorts, many of which were rancid or smashed flat by the time her flimsy cardboard boxes made it from Saint Francisville to Vietnam.

We read anything and everything, but at the time I favored classics, novels, and war histories. I ran across a quote from an Irish unit on the western front in WW1: If you are a soldier, you are either one at home or two on the front, if one you needn't worry, if two you are either one out of danger or two in it, if one you needn't worry, if two you are either one not hit, or two hit, if one you needn't worry, if two you are either one trivial or two dangerous, if one you needn't worry, if two you either live or die, if you live you needn't worry, and if you die, you can't worry! SO WHY WORRY?

LAME DUCK SHORT-TIMER

MARCH 5, 1966. *LCDR and BM2 staying with us while surveying island—for new Swift base?*

MARCH 5. *Present salary $690 month.*

MARCH 6. *Went skindiving alone via pirogue, no luck water murky.*

MARCH 9. *Rain, wind increasing to thirty knots.*

MARCH 10. *Admiral Ward here.*

MARCH 12. *Departed Binh Ba for Nha Trang arrived via Huey from Ba Thon.*

MARCH 15. *Toured provinces NE of Saigon w/Tuong and Colette in his Citroen. Dinner East of Saigon by lakeside-broiled crabs.*

MARCH 16. *Ate at Caravelle Hotel. (Delicious langoustines in sauce, I remember)*

MARCH 19. *Saigon. Saw King Bobby.*

I had ordered shoulder boards and silver clips showing my promotion up to lieutenant, but never received them, so for the remainder of my time in country I usually went around dressed as a lieutenant junior grade. I didn't care. I took some hops (scrounging a free flight on a military aircraft) to see the country. When we weren't patrolling or sleeping, I took days off to go to the MACV compound in Nha Trang and trade stories with other advisers. I could snag an empty room at one of the bachelor officer quarters (BOQ) hotels in Cholon, the Chinese district on the west side of Saigon. Bumming a flight to Saigon or most other sites in country was easy. In the hectic buildup, accountability and bureaucracy came last. I passed through Da Nang, Quin Nhon, Banh Me Theut, Dalat, and Special Forces outposts in the highlands flying in C-47, Caribou, CH-47 Chinooks, L-19, and Beaver airplanes, as well as the ubiquitous Hueys and H-13s helicopters. In the small II Corps airfields, fifty-five-gallon drums of Agent Orange would be stacked up right on the edge of the tarmac. On Caribou and Chinook flights, I probably escorted drums of the stuff while riding back in the cargo hold.

In the early 1960s into the buildup, Saigon was plagued by nearly daily terrorists attacks on restaurants and other gathering places by VC and sympathizers who would abandon bikes and cyclos (three-wheeled taxis, usually motorized) packed with plastic explosives on timed fuses—probably C-4 from dismantled Claymore mines. Still, Saigon was an exciting, smelly, noisy, diesel fume-filled madhouse crammed with every sort of humanity—troops, press corps, residents, remnant French colonials, refugees, orphans, street hawkers, sex workers, VC in constant motion. It was an R&R adventure in food, bars, movies, shopping, and trouble. Tu Do Street had become a half-mile strip of bars and food stalls.

As an old French colonial city, it had a scenic and romantic side, women in fluttering ao dais (silk pajamas with silk side-slit overskirts) along wide boulevards lined with green parks and nineteenth-century French architecture. Parochial school girls in tidy blue uniforms. But the scene was marred by every street being clogged with low clouds of blue diesel fume spit by cyclos, motorbikes, military vehicles, carts, bicycles, and a variety of French cars and trucks of every age in a perpetual noisy traffic jam. Most vehicles honked every minute. Across, on top of, and in between scurried people and dogs, usually carrying something on their back, head, or shoulders—perhaps groceries, a suitcase, or food to sell at a market—or hidden C-4 plastic explosive. The GI saying went, "We own the day but Charlie owns the night."

My counterpart and I had become good buddies, and in March he and his wife gave me a grand tour on my next trip to Saigon. I strapped my Browning pistol to my leg under my pants and we took off through the checkpoint into VC country in his Citroen, lunching at the French restaurant of a rubber plantation ex-colonial and sightseeing up country, northeast of Saigon. We slipped back into Saigon in the evening and hit some fancy spots frequented by locals.

LEAVING BINH BA

War would end if the dead could return.
    —Stanley Baldwin, three-time former
        prime minister of the United Kingdom

MARCH 26, 1966. *Last day at Binh Ba, LCDR Evrard arrived with junk division advisers by junk, had stuffed embassy chicken and French fries for supper, departed for JD 27.*

MARCH 28. *Departed Phan Rang (PR) C-123 for Nha Trang to pick up stuff.*

MARCH 29. *Returned to PR via Huey helo. PR off limits: Buddhist demonstration.*

APRIL 1. *Four persons detained patrol south to salt flats.*

APRIL 2. *Patrolled north to Vinh He. Patrolled to Cape.*

APRIL 5. *Surveillance flight Lt. Rainey.*

APRIL 6. *Patrolled alone down to Son Hai.*

Coastal Patrol Group 27. The Binh Ba base was squared away. In late March, they gave me a nice send-off and a commendation letter. I was transferred to the next base south at PR as the new SNA. It was a mess, only had one non-com adviser, and they worked to get rid of me from the get-go. The base was more vulnerable; it wasn't on an island and VC were active in the province. By this time, sandbags and concertina were more available and I picked up all we needed to fortify the base. Perhaps the VNN couldn't obtain them without me, but maybe they simply hadn't made the effort—the morale and discipline at this base was suspiciously poor.

I think many of the VNN sailors there were irregular recruits from local fishing villages with minimal military training, two or three pieces of khaki uniform, and entry-level subsistence pay. They were swapping a life of gutting fish and mending nets for armed camaraderie and adventure, and did the best they could with third-rate gear and chow. The few gung-ho ones sported a bravado tattoo, Sat Cong, meaning Kill Communists. The good news was that the US advisers were housed and fed a couple of miles off base in a US Army wood-framed, screened bunk house, and we had a jeep for transport back and forth. The bad news was we only got to know the VNN on patrol or when putting them to work improving the base defense. I only had a boatswain's mate petty officer first class (BM1) with me, so once again we were 50 percent undermanned.

I had to lean on my counterpart to get boats out on regular patrols; he usually didn't go out himself. From my pocket diary entries, the biggest fight was in February when we sank one junk, no bodies or material recovered. The entry four days later: *One VC body floated to surface this morn—searched area further—no more bodies—will patrol area tonight.* And two days later: *Good patrol last nite. Cdr. Toi kicked* ——*'s* [my counterpart, the base commander] *ass and he went out with me— detained two deserters, two VC suspects, four more bodies floated to surface.*

In hindsight, Base 27 was much more vulnerable to attack than my old base, and I was the expendable short-timer adviser on his way home. I suspected that my counterpart was so hostile and uncooperative because he might be a VC sympathizer, a fox in the hen house. Maybe I will go back and research how quickly they were overrun in later years by the enemy as compared to better commanded junk bases.

APRIL 7, 1966. *Joint ops, blocking 0300–0730, five suspects. 1930 two junks est. twenty VC.*

APRIL 10. *Took mortar and bazooka to Cana drove one co. VC from village.*

APRIL 13. *Three junks to Son Hai as reaction force against VC in village.*

APRIL 14. *Surveillance flight pilot James.*

APRIL 15. *Surveillance flt, pilot Rainey.*

APRIL 16. *Patrolled to Vinh He.*

APRIL 18. *Surveillance flight with Rainey 1330, abandoned village north of Son Hai—amphibious assault 1630–2030. N.B. Forrest—"Keep the skeer on em."*

SCROUNGING FOR MORE FIREPOWER

I PICKED UP TWO more 60-mm mortars for the junks, which we put on the bows in a semicircle of sandbags. I also got a bazooka and a .50-caliber machine gun, which I later tried out on a VC coastal village. Fortunately, the operation didn't last long because I was firing from only 200 meters or so out in the surf. I don't remember calling in fire support. A spotter plane reported that we drove out a company of VC. I reckon that an attack from the sea with the lead boat firing 50-caliber shells was an ominous start. Hitting a bullseye to blow up the shoreline bunker from just outside the surf was enough warning of impending disaster. They cleared out to the east.

Unfortunately, the .50 caliber was an Army Browning on a tripod, not the stanchion-mount Navy version with metal shield. I was firing it without realizing how much louder it was than the '30s. I had ringing ears for days and eventually ended up with significant ear damage as documented in my mustering out medical exam in Saigon later that year. I fired one round with the bazooka, a bullseye on a concrete bunker sitting just behind the beach in the palm trees we were taking fire from. That was the one and only bazooka round I ever fired, so I was 100 percent on target with bazookas in Vietnam. I always wondered where that bazooka ended up being used and by whom.

Navy people untrained with Army weapons meant we spent a lot of time reinventing the wheel. Based on my experience, firing a 60-mm tripod mortar on a rolling junk was a fruitless exercise. We finally took them off the junks and gave them to our counterpart for defense of the junk base. Except for perimeter defense, ARVN and US grunts had probably gotten tired of lugging them around the country in this nasty guerrilla war of wandering patrols, so I reckon they were widely available for any takers. I think some Swift boats and Riverine craft carried 81-mm mortars, much more robust and effective weapons.

APRIL 21, 1966. *Bob York from ARPA* [definition not known] *Saigon visiting to photograph junks for Junk Blue Book.*

APRIL 22. *Picked up eight suspects from slave driver papa.*

APRIL 23. *Three deserters: one man improper papers, three military age, patrolling south to Cana with CO.*

APRIL 28. *Patrol to Ca Na 2000-0615, 4 detained.*

APRIL 29. *Patrol with* [Swift boat] *46.*

APRIL 30. *VC hit Pho Tho BN 848757. Swifts fired eleven rounds 81 mm.*

MAY 2. *Patrol to five miles of Cana, Swift 50 in area, no suspects.*

MAY 3-4. *Heavy afternoon rains.*

MAY 5. *Burned-up generator.*

MAY 6. *Stock market took worst loss since Kennedy's death.* [How did I find this out?]

MAY 12. *Patrolled south to Son Hai. Checked about twenty-five boats, three suspects.*

MAY 27. *Sixty days to go. "Thank Him for his kind intentions, go and pour them down the sink, if an angel out* *of heaven, gives you something else to drink."* [Source: G. K. Chesterton, author and philosopher.]

NEW PATROLLING
MUSCLE ARRIVES

Do not needlessly endanger your lives until I give you the signal.

—President Dwight D. Eisenhower at the German front in France

SHORTLY THEREAFTER, SWIFT BOATS (50-foot patrol craft fast [PCFs]) were assigned to II Corps surveillance duty—they were noisy, speedier, and more aggressive than wooden junks! I was contacted to accompany the boats on their first patrols from Phan Thiet north to Nha Trang, and I showed them navigation points, suspect villages, and coastal geography. After months on wood junks, I was wowed by the Swifts' speed, roar, and firepower, like going from Trek bikes to Harleys. They had either twin 50s or 40 mms (with a metal shield!). They must have been fitted out with 81-mm mortar also, because my diary says VC hit Phu Tho on April 30 and Swifts fired eleven rounds 81 mm. By the end of my year, the Swifts and other US units were integrated into the coastal patrol operation, now designated Market Time, and coordinated through the surveillance centers.

In my corner of the war, battlefield heroics were the stuff of bar gossip in Saigon or somebody else's base camp. I knew nearly nothing about this facet of war because my year in that theater of combat was spent mostly with Vietnamese people, with only a handful of Americans on hand. What happened in firefights was personal and people generally seemed to do their assigned jobs. Nobody was around to hand out medals, slap backs, or toast victories. The last patrol was mostly forgotten after our next one was completed.

Twice a week I would make coastal surveillance flights in an L-19 single-engine, two-seater spotter plane. I took along grenades and my AR-15 for backseat offense as we had an open cockpit with only two windshields. We figured anything was a plus because we carried only two wing rocket pods and smoke grenades. These planes were known as spotters. I would fly back seat twice a week up and down the coast ready to assist surveillance or whatever other support was needed. The pilot let me do some touch-and-go's to make sure I could land the damn thing in a pinch. We were mostly at loose ends until directed to observe fire or locate targets. Our flights were in the morning within ten miles of the South China Sea coast. We could see the ocean and steer visually by paralleling the water unless we were flying lower than 500 feet or so. I found out decades later that the Air Force sprayed Agent Orange at night in sections of the jungle we flew over in the morning. We probably flew through clouds of the stuff.

We took holes in the wings but you could never see or hear return fire wearing headphones. The most memorable incident firing the rockets was when we spotted a pod of orcas over shallow white sand in a bay just off a fishing village. Good Samaritans trump sportsmen and conservationists, and I can confirm that on the return trip up the coast, half the village had manned boats and tow ropes for the seafood harvest—we would have loved to find out they had a feast, but weren't sure which team they were on. Luckily, we took no fire flying over at 300 feet when we wigwagged them.

JUNE 9, 1966. *Leave Saigon for Hong Kong* [my R&R week]. *"Today's a gift, that's why it's called the present.* [Source unknown]*"*

JUNE 9. *Spending* [HK7$ = US1$]: *tie tack HK$35, wallet HK$95, ring HK$236, suit HK$222, two pairs shoes HK$136, six shirts HK$124.*

JUNE 9. *My birthday. Spent four-and-one-half hours in the goddamn Hilton Dragon Boat Bar getting polluted with two ex-Navy and a Huey pilot from Dong Ba Thuan.*

JUNE 9. *Picked up my six shirts—wrong pattern. Arrived Saigon 1415, spent US$348 total in Hong Kong. Arrived Nha Trang via Da Nang, Pleiku, Cam Ranh, Quy Nhon route.*

R&R

EVERY PERMANENTLY STATIONED MILITARY man in the war zone for a full year was given one week of R&R. In lieu of leave, that was the standard earned vacation days for those on year's duty in a combat zone. In 1966, our R&R choices were Hong Kong, Bangkok, or Tokyo. (Later, Sydney and perhaps Manila were added.) I put in for Hong Kong and my choice was approved.

As my diary implies, it was mostly drinking, eating, sleeping in a real hotel bed, and a little shopping. I don't remember as much about the food and dining as I did in the weeks when I returned after leaving Vietnam for good. The second visit, I pigged out. I could hardly focus on anything but filling my stomach my first days there.

On my return to Saigon, I went out to dinner at International House with my Irish officer friend who was the admiral's aide. I gave him my .25-caliber Browning boot pistol when I left Saigon, thinking this poor bastard was going to need it more than me. I probably could have sold that neat little non-military issue semi-auto at the hotel or the Special Forces compound for several hundred dollars—a lot of money at that time. I was to run into my Navy buddy again in Thailand and he returned the favor.

JUNE 19, 1966. *Dow Jones average 897.*

JULY 8. *Infected ear.*

JULY 9. *"Religion is more poetry-plus, not science-minus.* [Source unknown].*"*

JULY 25. *Big gripe: disposition of prisoners.*

AUGUST 5. *Purchased exit visa. Verbal orders from Cdr. Reagan in Saigon to get released!*

AUGUST 5. *DISCHARGE DAY!*

That which is past and gone is irrevocable, and wise men have enough to do with things present and to come.
> —Sir Francis Bacon, English philosopher
> and statesman

IN THE HOME STRETCH

MY YEAR IN COUNTRY wound down with a few weeks at the surveillance center in Nha Trang. I was more than ready to wrap up my contribution to what was looking like a war where the alleged losers would suffer mostly negative news and its recounted history would need to be skewed and politically corrected to sell. Later, it hit home that I had spent much of my year without the camaraderie of fellow American combatants, essentially alone. Being in combat messes with your mind. It made me even more self-reliant, but also a more insular person. Any idea of what they used to call shell shock in earlier wars and now call post-traumatic stress disorder (PTSD) never occurred to me. While seeing some action, I was not in harm's way as frequently as others; my year was mostly a nervous grind with lots of waiting and boredom between occasional bouts of scary stuff.

> *And the truth is that all veterans pay with their lives. Some pay all at once, while others pay over a lifetime.*
> —J. M. Storm, author

I had earlier requested to be detached in Saigon in order to tour Asia. I caught a Caribou to Saigon, debriefed, got my physical, gave my .25 semi-auto pocket pistol to my buddy, saluted military life goodbye, bought a Hawaiian shirt, flew out a civilian on a purchased airline ticket back to Hong Kong, and went to the Hong Kong Hilton. I went on to Japan and the South Pacific, and then worked in Australia for a year—backpacking across Asia and the Mideast and living in Paris. Like a lot of early vets, I scored a cultural goose egg totally bypassing the Great American-Janis Joplin-Grace Slick druggie hippie burn-down-stuff-and-riot revolution that had taken hold in the States.

WANDERING POST-VIETNAM

MY LITTLE 1966 POCKET diary was full of entries by August 7, 1966, where I drew a big star and entered: discharge day. Happy times! The week before I had purchased tickets on Thai Air and purchased an exit visa from South Vietnam to leave as a civilian.

Entries on previous page: *Books recommended*—The Last Gentleman, *Walker Percy;* Earthly Paradise, *Colette;* Ariel, *Sylvia Plath;* Remembrance of Things Past, *Marcel Proust.*

My literary ambitions remained intact and my literary tastes were fixed at a loftier level than what evolved later in life. My idea was that music and art are self-expressive, but literature is more so. The best works bare the artist's soul and perhaps their true vision of life. Literature is more responsive to personal ambition. I had absolutely no musical ability and only raw artistic talent, so always looked to books.

> **AUGUST 8, 1966.** *Maybe I should find a good university in Australia and request to audit some lectures.* [I was already planning to reach Australia but must have been mulling over what to do whenever I got there. Just living day to day, war behind me.]
>
> **AUGUST 10.** *Departing Saigon for Hong Kong/Tokyo 1405.*
>
> **AUGUST 11.** *Spent nine days in Hong Kong.* [The first day I took a tour bus around New Territory and Hong Kong

Island and ended up with the tour guide, Ella, swimming in Repulse Bay, going to discotheques and floating restaurants—relaxed and moving on. The nine days cost me HK$447. I flew on to Tokyo.]

I had a memorable first night meal at the Hong Kong Hilton restaurant. On the menu, I started at the top where Australian rock oysters were first listed under appetizers, thinking "I'm going to eat every damn thing on the menu." Of course that was impossible, but I have forever after enjoyed oysters on the half-shell after disliking them since childhood.

One morning I took the bus tour up to the China border. This was the closest I got to Taiwan or to mainland China, which was off limits in those days. My cousin Bob Barrow (General Robert H. Barrow, 27th Commandant of the Marine Corps) had headed a team conducting guerrilla operations in WWII destroying railways and bridges in Japanese-occupied central China. When talking with Madame Chiang Kai-shek (also known as Soong Mei-ling), who was a prominent promoter of Republic of China affairs, he was asked about his fluency in Chinese dialects. After responding that it was limited, she said that the more fluent in their language the more he would think like a Chinese—and that was bad! Madame Chiang was a graduate of Wellesley College and fluent in several languages.

Bob was serving in Vietnam after I was back and working on Wall Street. I happened to see him interviewed on national TV by Dan Rather during the battle of Khe Sanh. Years later, I took my family to the Pentagon to visit him when he was commandant. I was there to run the Marine Corps Marathon—in miserable, snow-driven, near-zero weather—and got my certificate of completion signed by him as a consolation for the semi-fun weekend. (I missed the qualifying time for the Boston Marathon but qualified the next year.)

I spent much of my Japan visit in and around Tokyo, commuting by fast train forty minutes or so back to the US naval station at Yokohama where I could stay for free at the BOQ or with Gary Lintz or other friends living in or stationed in the area. I got on Gary's motorcycle—my first time on a motorcycle. I think Gary had been in the Navy and stayed on in Japan working for IBM. He gave me a few pointers on shifting, and I eased into the road. After a few hundred yards and a touchy turn or two, I soon had enough of riding down the gravel roads near his house.

I took the bullet train west to Kyoto, an ancient cultural center in Japan, staying in Japanese inns called ryokans, eating clear soups, sticky rice (bland meals for my Cajun tastebuds), dining cross-legged on futons.

> *A man should share the action and passion of his times at peril of being judged not to have lived.*
> —Oliver Wendell Holmes, former
> Supreme Court justice

By now, I had acquired another notebook small enough for my backpack in which to keep more elaborate notes. *Just watched an evening cartoon show on TV in my ryokan. The show was WWII Pacific theme, Japanese planes battling enemy planes. Obviously Yank pilots by the facial caricatures. Yank planes severely dumped and destroyed by both Japanese planes and nature—in one scene a huge flock of ravens fly over the Yank planes and defecate on the planes' windshields! Show ends with paratroop planes braving the air war to chute down to an oilfield-laden island.* Japan was always pinched for petroleum.

I mulled over my one hostile exchange during my trip with a countryman. *A young bartender kept bringing up the Hiroshima A-bombing. I agonized over finally meeting two interesting*

Japanese women my last night in Japan, and getting dumped by one late at night because of my own timidity. [Going out with another, I was taken aback by her blatant attempts to have me buy her western clothes at a department store, yet pleased at myself for wiggling free with my wallet intact.] *My love (or let's find some human to talk to!) life has fallen victim to the vicissitudes of cultural differences and the hangover of wartime animosity, but also to my awkwardness returning to civilian society.*

I went shopping for watercolors unsuccessfully after putting a reminder in my notebook weeks earlier. I enjoyed drawing, and painted with oils in high school. A swamp painting I did was hanging in a café in Saint Francisville. I bought some small watercolors while on leave in Vietnam, and now that I had returned to the real world, I was again stimulated by art and museums.

I became fascinated by Japanese cultural differences. What makes the Japanese culture great?

1. Highly moral environment (cleanliness, honesty, courtesy)
2. Hard, disciplined work habits [Toyota and Lexus are more dependable cars]
3. Sense of artistic and social proportion (economic rationalism and refined creativity in architecture, interior decorating, craftsmanship, modern art, and related skills—arts)

Their uniqueness is obviously the product of eons of cultural isolation with sparse input from gaijin (foreigners). Bottom line, they were ill-prepared to deal with outside influences.

SEPTEMBER 3, 1966. *I eat wasabi with everything. (I miss my spicy foods!) Japanese babies are badly spoiled. It has been said that Japan is a children's paradise. Japanese* *words mostly end with a vowel. Primary consonants, descending: KTMYHRZSPBNCW. Rode train up to Takarazuka for the girls' revue, a wonderful, highly professional and lively Broadway-style musical. Also saw a semi-classical drama with great singing. One of the highlights of my trip.*

I was reading *The Scope of Fiction* by Cleanth Brooks, fueling my writing ambitions. With all my fiction reading in the boring parts of combat duty, I should have noted that some of the most famous writers seem to be conflicted drunks short of money with sundry personal and social problems and that thankfully I wasn't following that script. But what made me still think I could be a successful fiction writer?

SEPTEMBER 9, 1966. [Notes on travel costs.] *Japan to Australia by plane $476, Sydney to Naples by ship thirty days for $482—Athens, Auckland, Europe via Africa-Suez, Bombay, Port Said, Athens.*

My wandering around Japan was ending and I cataloged the peculiarities of the place: *Facial cleansers in restaurants—all the pimples! Umbrella locks—but nobody steals here. Sliding doors everywhere, stainless steel dinner plates, red phones, disposable toothbrushes, crappy garment materials.*

I was back in Tokyo and Yokohama by September 10, then headed to Kyoto for eight days before leaving for Kobe and Osaka. I took the high-speed bullet train round trip to Kyoto. Until 1868, Kyoto was the imperial capital of Japan, which accounts for its cultural prominence and historical importance in the country. The Meiji Restoration marked great political and social change as the country moved toward western ways and entered the Industrial Revolution. The imperial capital shifted to Edo, and its name was changed to Tokyo, literally eastern capital.

When WWII ended, General Douglas MacArthur oversaw the occupation of Japan from Tokyo, where he was the effective ruler even as Hirohito continued to occupy the imperial palace. My father-in-law, General Douglas Kendrick, accompanied MacArthur's staff, serving as head physician. He participated in a curious duck hunt. The palace maintained several ponds around Tokyo attracting wild ducks by the thousands. With the aid of clipped-wing "house" ducks and liberally sprayed grain, the emperor's guests lined up along canals radiating like spokes from the bamboo-walled ponds with long-handled butterfly nets to net the wild ducks as they flew down the canals. Lubricated with saki after a leisurely lunch, guests often ended in the canal instead of the ducks ending up in their net. Dr. Kendrick later left me two Samurai swords and a silver incense container presented to him by the emperor. He had treated the emperor for hypertension and performed an appendectomy on his daughter.

I kept lots of notes on the American stock market, which I could now follow more closely. Also, I kept notes on basic Japanese words and phrases, which I mostly picked up from sitting on bar and music club stools and the few girls and guys I met who knew some English. The Japanese were still reluctant to communicate with outsiders at that time. Younger Japanese, now twenty years post-Hiroshima, in sympathy with their parents and reverential of the dead, may have felt it a sign of disloyalty to talk to me.

> **SEPTEMBER 29, 1966.** *I find that true politeness consists of manifested innate goodwill toward others. By this definition I find the Japanese not truly a courteous people. Rather, their society calls for certain well-defined courtesy*

*rituals—established habits or modes. While window-
shopping, Japanese individuals will step in front of me to
gaze at something I was busy viewing. Japanese motorists
are extremely impolite as are streetcar and train passengers.
While socially progressive in nearly all areas, Japanese
formality, conservatism, and, indeed, inhibition manifests
itself in public gathering places. Dancing is not very
popular. I have yet to find any dancing spot in Kyoto, and
the few places in Tokyo showed me how terrible they are at
dancing, despite the gracefulness and rhythm of geishas and
maikos.*

My time in the country was ending, the latent hostility I experienced may have made a bigger impression than the posi-
tive experiences. Traveling alone without any fluency made it
harder going, and a young western male was probably assumed
to be a touring ex-soldier Yank in those times. I was a ready target at times, and my diary detailed encounters where I nearly
got led astray on shopping forays with deceptive affection and
such. *I am so skeptical of Japanese girls that I had no trouble keeping a light humor . . . .*

OCTOBER 2, 1966. [I bought a ticket on the Nanchang,
a Japanese freighter.] *Went shopping with New Zealand
couple and split fare to the Toyo department store. Spent
most of the seven hours ashore at Osaka Castle, built in
the sixteenth century by the shogun conqueror of Japan.
Magnificent scale and quite interesting study of primitive
medieval military fortification.*

OCTOBER 4. *Underway at last, at 2:30 p.m. I thought
as I watched us moving out the harbor, how ironic to be a
paying passenger on the ocean less than two months after
the completion of a four-year period when they were paying*

*me to go to sea! Wouldn't my old Navy friends be curious to
see me now! New Zealand term: to grizzle (complain).*

OCTOBER 5. *We are carrying a load of steel for the
Fiji Islands on the way to Auckland, New Zealand, an
uneventful voyage of twelve days in calm Pacific waters.* [I
was one of five passengers on board, and we took all our
meals at the captain's table with the two to three other
officers.] *The food was excellent, and the only English
spoken was by the captain. Boarded Nanchang late this
afternoon at Osaka Harbor. My first impression very good.
Comfortable stateroom, drink prices very cheap (quart
of beer HK$1.50 or US twenty-seven cents). Lounge is
nice, dinner very decent, and western stewards civil and
deferential, guests cosmopolitan, captain standard-British
type on first impression. Indeed, we must wear jackets for
dinner—I was surprised.*

*Ship is in freshly painted condition and looks very clean
for its type. Cargo mostly steel. Am listening to Beethoven
symphony on my little Sony portable with antenna
protruding from the porthole. A pleasant little joy, wearing
my yukata* [Japanese robe] *and having an Asahi beer.*

OCTOBER 7. *Continuing to Fiji Islands. Seas very calm all
the way and hope they remain so. I had never been seasick
in four years of sea duty in the Navy, but had gotten very
sick at thirteen on Uncle Francis's 48-foot fishing boat in
the tail of a hurricane when we nearly didn't come through
whole. Strangely, years later I sometimes got nauseous when
out grouper fishing in choppy seas although by then I had
worsening ear issues.*

I studied the two passenger couples on board, British
shipmaster-Japanese wife and Auckland tourists. *Good
blowers in cabin, nice breeze shooting through porthole this*

morning. Approaching equator and there are squalls here and there. We have a portable saltwater-filled swimming pool—canvas and steel—on main deck forward. Delightful to get a tan once again. Last tan was at Kamakura, Repulse Bay—slight pang for Ella's company. I got excited about the tropical Fijis. I was crazy about tropical isles, mostly due to my Caribbean experiences and reading—Joseph Conrad, Robert Louis Stevenson, Graham Greene. I spent my first year as a naval officer in Cuba, Puerto Rico, Virgin Isles, Jamaica, and the Keys in the Cuban crisis period 1962–1963.

Meals on the Nanchang are a curiosity to me. First, a typed menu on a card is presented prior to each course, invariably a soup, fish, then a main course with vegetable choices listed, then dessert or cheese. I suppose this is the English sea-going style. Fish knife and fork. Second, the variety of dishes, tonight flounder followed by quail. At noon, curry among other choices, yesterday whole rainbow trout. Drinks before supper, then a toast to whatever—I can't recall. I doubt to the warmonger, yours truly.

OCTOBER 16, 1966. *Arrived Suva, Fiji. In early morning took a quick walk around the town. While in Fiji, I met an unmarried young English couple backpacking the world like me, and we traveled together for several days. They were attractive-looking, devoted—clinging—to each other, most certainly working-class Brits with obvious socialist sympathies; we had our differences after they found out I was a university-educated naval officer.*

I had never experienced such class resentment from the get-go, even back in Louisiana. Our conversations were constrained within certain bounds, so I never found out anything about their parents or their schooling and religion. They reciprocated, asking little about me. Still, I always accepted people as they were—whatever works, whatever the situation. It was in the afternoon that I met them, Rick and Sally, while watching a game of bowls, lying in the grass on the side line.

I got off the Nanchang, terminating my voyage, with a thirty-pound refund in order to join them in a cruise around the smaller Fiji Islands—three pounds for six days! We went swimming in Nabutini, then went bike riding. Sent postcards home to Mary [sister- or friend Mary in Saint Francisville?], *to my old Chattanooga girlfriend teaching at Virginia Beach, and Annegret* [German girl from Paris trip].

OCTOBER 17, 1966. *Still waiting for offshore weather to break. Visited Fiji Beer Factory, free beer, introduced around at Yacht Club by the "commodore's wife" a jovial, tanked, red-faced portly woman at the bar. Bus trip to Nadroumai, sits on a big river. Had tea there. Big picture of the Queen in shop. One street town. Came back by bus. Fooled around town, giving up on the weather getting better.*

The next day we went to a coastal fishing village and I stayed in a little thatch hut with a picture of Queen Elizabeth above my bed. Wonderful, friendly hosts. I remember the village party drinking kava brew one night—lasting past midnight: loud, music, dancing, the kava, a muddy-looking and tasting mix from local roots reputed to be an aphrodisiac. We stayed several days.

The villagers subsisted on seafood, gardening, and coconuts, kept a stock of captured sea turtles turned upside down behind the huts, and every year some of the villagers got work permits to labor in New Zealand for cash to buy pots, spoons, fish nets, portable radios, clothes, and other necessities. I soon took a work boat, a small ship used to ferry workers (some permit workers from Fiji) island to island. I recall they could get a work permit once every two years to do manual road work and such for six months, during which they accumulated and sent home (and maybe drank up some) enough cash to buy subsistence needs like pots and pans for their return.

OCTOBER 19, 1966. *Made reservations on the Oriental Queen* [What a name! The Queen turned out to be a rusty mostly open deck work boat]. *We rented bicycles for two shillings/hour and rode around the outskirts of Suva along the beach. A big cruise ship, the Iberia, in town, P&O Lines. Had picnic along seawall.* [I was with the Brits.] *Laboriously (!) cycled around behind Suva and then to the yacht club for beers and salted peas. Either I was a rusty bike rider, the bikes were lousy, or probably both. The following year I got on a bike in Bali and nearly fell off from awkwardness—my recollection that Bali was my first bike ride since childhood obviously a mistake.*

Came back and went swimming in Suva at the sea baths. Watched the Pacific Games swimming trials—a tiny eleven-year-old Indian girl named Olive Pickering won all the girls races.

OCTOBER 20, 1966. *Went out to Nabutini—sixty miles by bus, two hours and forty-five minutes along the sea— with Rick and Sally. Hired a boy with a canoe to pole us across the bay to a lovely white sand beach with crystalline waters. We hiked the beach around the bend to the outside facing the barrier reef where Frank Ransom's bures (huts) were situated. "Boola" is Fuji for "hello". Ransom wasn't there but a nice Australian lady, a guest in one of the bures, took us in and fed us lunch of leftover curry. Later swam,* *shell-hunted, and napped on the beach. Found and opened a coconut.*

Later, we headed back to the village in order to catch the 6 p.m. bus. Bob Ryan was a middle-aged white, disabled Kenyan who left Kenya because of the increased cattle theft, taking this cottage and living a hermit's life, a pleasant English fellow, eager to talk with Sally and Rick. Ryan invited us in where we inquired about crossing the mud flats at low tide. Had a brother in New Zealand and seemed content. We started across the mud flats, myself in the lead, mud up to our knees in places, three-quarters of a mile wide. Finally noticed a Fijian walking quickly along a shallow ditch; we followed him and made easy progress. Listened to the radio and slept in the bus on the way back.

OCTOBER 21. *Arranged tickets to New Zealand-Australia. From noon to 3 p.m., Sally, Rick, and I drank beer in an open street-corner bar, trying to spend the ten pounds each of us had saved by exchanging air ticket for sea ticket. That night I ventured to Suva's only cabaret, open until 2 a.m., the Golden Crown (otherwise, drinking ceases at 9 p.m.). A motley crowd, Teddy boys with high collars and white bucks, ugly Fijian and Indian sex workers, and odds and ends from the waterfront. I soon departed.*

OCTOBER 22. *Today marks two years without smoking a cigarette.* [I never smoked another cigarette again. I quit smoking for good one morning after waking up with a hacking cough while in port at Charlotte Amalie, Virgin Islands.] *Returned by bus to Nabutini, being poled across by Frank where we had spent the afternoon. Took lodging in a spare one-roomed house with two beds. Watched a local rugby game. Had supper of turtle liver* [villagers had killed a big sea turtle that morning], *tapioca, taro root, and tea.*

Later we bought some Fiji grog—kava—said by Frank to be an anti-aphrodisiac. It comes dried in brown packets of one shilling each. Our charming Fiji host and hostess mixed it up and we had a little drinking ceremony, the stuff mixed with water and strained through gauze in a two-gallon wash basin. Hand you a coconut shell full, everyone claps three times, drink in one swoop, then repeat with the next guest. Meanwhile, two boys playing Fiji songs on guitars, and one with a ukulele. Others singing along. There were embarrassing pauses of anticipation between numbers as Fijians waited for us to get up and do some wild western dance.

Finally they did a couple of dances themselves, then we all joined in mostly doing a Fiji circle dance. Seems that the last visitors had stayed all night twisting and the same was expected of us. Slept under straw mats in bed with a big picture of Queen Elizabeth looking down from the wall above the headpost. Awoke at dawn with sand in my hair and salt on my body, unshaven—and growing a mustache. Took a swim at the stunningly quiet beach.

Later, the native drum-beating woke up Sally and Rick. In the morning, we walked back across the mud flat and spent the day in the sun. Too rough to go out in Frank's boat, but we borrowed his skin-diving gear and drifted along the shore in the tidal current looking at tropical fish in the coral-studded sand. Later, Frank showed us his shell collection, we had coffee and talked about sharks, then he drove us back to the highway in his Land Rover. The journey by bus back to Suva seemed much quicker this time.

OCTOBER 24. *Waiting for Oriental Queen to sail at 8 p.m. What a name for a rusty workboat! Had picnic in Suva park. Got underway on time amid great fanfare—band*

*on the pier, colored streamers, yelling crowd of about 500.
The passengers were pretty evenly split up between Aussies,
Kiwis, Indians, and Fijians. The officers were Japanese,
crew Chinese except one—the only US crewman on board.
Four-man cabin: two Aussies, named Ray and Nicholas,
with me; Sarah in dormitory. Picked up Aussie girl on
dance floor first night named June—okay, but no place to go
so we separated at 2:30 a.m. Generally spent my evenings
drinking on the dance floor with Rick and Sally.*

*One evening they had a hat contest and I made a funny hat
but didn't win a prize. We stopped at Vavau, Tonga, on the
way over—a pretty little village. We walked out of town
and down through the fields of a coconut and pineapple
plantation. Later, a truck gave us a free lift to a coastal
village about two miles further down the road. We had
stand-bought tomatoes, papayas, and a coconut for lunch.
The natives were very friendly ("Friendly Islands").*

Ceremony for our ship at King of Tonga's house

OCTOBER 30. *Arrived Auckland at 8 a.m. on a gray chilly morning. Took taxi to a big old boarding house at 2 Basset Street. Spent the day out on yachts planning to sail over to Australia. My English chums were rather antagonistic, and I in return. About 11 a.m., Sally said I should strike out on my own because I didn't seem to be meeting people, etc. I complied, icily. I was thinking exactly the same thing, getting stuck with them was simply laziness on my part and was keeping me from meeting more lively people. Rick was mousy, acquiescent, and quietly complicit. Sally resented that I refused to make a play for her? "Hell hath no fury . . " She struck a nerve about my meeting people, though—may the bird-of-paradise fly up her nose. Meeting people—any band musician can tell you—first, you smile at them. Guess I was either too shy or an antique Southern gentleman type who would rarely rise to the occasional come-on.*

OCTOBER 31. *Sally and Rick took off for the interior, first being very polite and giving me contact addresses in England. I still feel a little nasty about them though. Kiwis don't like Poms. This name is wrongly believed to mean Prisoners of Mother England, but primarily refers to "them" of an "us versus them" situation and a contraction for people shipped out from potato land (pomme de terre). It is also the Aboriginal nickname for visiting Brits sunburned the color of a pomegranate fruit.*

NOVEMBER 1, 1966. *Left Auckland at 9 a.m. hitchhiking out the motorway. Beautiful country, two truck rides right away. Lunch in Hamilton at a fellow's house who had five kids, a very nice guy. Had traveled in Europe and the States. Stopped at Lake Taupo, bought food. Got picked up by a fourth-generation Kiwi lady who hated Poms and advised me to pin my little Yankee flag on my shirt pocket. Was then picked up at the edge of Lake Taupo by an aircraft/*

sportscar man in "the only Lincoln Continental in New Zealand." Two cute Aussie girls inside. They soon got off at the youth hostel, though. I continued on with him through Desert Road past the beautiful snow-capped volcanoes and mountains. He left me off in Taihape after a delightful talk; as evening approached, I got picked up by a kid in a red pickup who took me all the way to a coastal town where I spent the night. He had just returned from Vancouver.

Spent night in a guest house. Ate in my room—kippers, cheese, bread, half bottle of New Zealand sauterne (too sweet). [At 26, I probably hadn't drunk enough types of wine to know how cheap sauterne's sweetness would shock my taste buds.] *Total first day's mileage: 350.*

NOVEMBER 2. *Headed for Wellington at 8 a.m., crossing the town bridge and riding with a traveling salesman, Schlage Locks. Later, I ran into a cocky, obnoxious blond stout Californian with a pack and bedroll who hitchhiked awhile with me. He claimed to have traveled eighteen months in South America, then joined a US research ship to Antarctica. Four months in New Zealand; threw some half-ass Spanish at me. Only the second American I have met so far (since Japan).*

Arrived Wellington at noon and Hans from Port Line, Ltd. gave me my Merrill Lynch check. [I don't remember if I had any bank account but my little stock account must have been at Merrill—likely I was arranging to pick up travel money at future destinations.] *Walked around town until 8 p.m. ferry left for Christchurch. Hans is an ex-Indonesian Dutch colonial type.*

NOVEMBER 3. *Arrived Lyttelton 7 p.m. Rode a horribly slow train into Christchurch. Began hitching at 8 a.m. One ride with a dairy truck driver—we made about ten* *deliveries. Heading toward Canterbury Plains—sheep country. In the afternoon, I got stuck in a chilly crossing for an hour and a half, northwester blowing. Later was dropped off by a farmer at Lindis Pass, a barren desolate valley, cold and with a high wind blowing. In the night, it frosted. I got down off the road, behind a clump of bush wearing my green jacket over my sweater and feeling fairly miserable. A big bull across the fence twenty yards away watched me, a magpie cawing and diving at my head.*

Finally got a lift with Matt, a nice sheep farmer who was bringing some schoolchildren from Christchurch for a week on Tarras farms. I remember when jumping in his car, asking him where everyone had gone—the road was almost void of traffic and deserted for hours. He said, "You haven't heard? We just got TV service out in the country today for the first time! If they had already bought a telly, they were staring at the thing smiling."

Then we had a flat tire one hundred yards down the road. I helped him with the change. Stayed at Matt's farm, rabbit hunting in the evening by car. [One of my clearest memories of New Zealand, when he found out I was a Vietnam combat vet and Louisiana hunter, was Matt sitting me on the front-right fender of his beat up old field car with a single-shot 22. Deathly cold there. I got two for four shots. I had remembered rabbits everywhere in the headlights, a Kiwi bunny plague—not!] *Unheated house. Too cold in the morning to take bath. Big breakfast and then Matt took me out to the main road. Offered to pay him for the stay and meal but no go.*

NOVEMBER 4. *By noon arrived in Queenstown, a resort beside a lake. Beautiful mountains, the snow-capped Remarkables, in the background across the lake. Soon* *found a guesthouse run by a nice Scotch woman. In the afternoon, took a bus trip to Coronet Peak and road up to the top in a ski lift.* [It was probably my second time on a ski lift, the first time was at thirteen on a trip out west when we rode up an Aspen lift in summertime.] *Lightly snowing at the summit. Rode down the mountain silently, only the voices of myself and the Aussie girl I rode with. She was a quiet, delicate, refined blond, a little shy. Regretfully, I said goodbye at her hotel after taking the trip back talking to her. I didn't see her later in the evening, though I looked. At night, everyone in Queenstown walked about or rode, bored—I too—until about 10 p.m. when I gave up and retired.*

NOVEMBER 5. *Got two rides—up to Cromwell at 10 a.m. Then, it started raining and I had no luck hitchhiking so I took the 10:45 bus, riding it through the Haast Pass in the rain all the way to Fox and Franz Josef glaciers. Motor-camped at the glacier. Caught a ride up the three-mile road to see the dirty-blue shadowed white glacier itself. It's moving nine feet per day. Torn path of slick rocks and water. Lush semi-tropical vegetation in this part of South Island—always seems to be drizzling here. That night I cooked my rations (kipper snack, beans, cheese, bread, pop) and talked with four Aussie girls hitchhiking around the country.*

NOVEMBER 6. *Left Franz Josef walking one-and-a-half miles before hitching a good ride with three men, one a very inquisitive broadcaster, and getting all the way to Greymouth. Sunday morning but we found a pub at noon and I walked out high. In Greymouth, more rain. I walked across the bridge, watching men fish for whitebait with big dip nets. Stopped for lunch at blowholes and pancake rocks on the Tasman Sea's coast. Beautiful coastline and* *lush flora. Walked two miles from there—a hole in both shoes—picked up by a Maori tribal native and a boy. Made Westport by evening—a typical dreary dirty coal and cement coastal town of 5,000 people. Raining, hole in wet shoe. Went to a western with the son of the guesthouse owner, a local newspaper reporter. He was skeptical of the benefits of travel, never having even gone to North Island in his lifetime. Later, he confessed to having doubts about staying in Westport and working on the newspaper; sadly, wondering if he should get a job at the cement(!) plant where he could make twice as much.*

NOVEMBER 7. *Ride with a fruit and vegetable delivery truck. Boy had captured a rare gray two-foot-long kiwi bird. Ride with manager of the state coal mines. Took me around some 3,000 foot coal mines. Next rode with the young geologist of my first day on South Island who previously was headed for Mount Cook. Then rode with an elderly horse breeder who drove very slowly. Arrived in Christchurch at 5:55 p.m. Evening at a dumpy guesthouse. Scottish landlady, weather very cool. Walked around the pretty but very dead town, cops in fin de siècle white hats and double-breasted suits. Very English flavor everywhere, schoolgirl uniforms, boys in black suits and boaters. Good-looking girls in the streets.*

NOVEMBER 8. *Woke up late, sunken in a hollow mattress. Cool, rainy, it's New Zealand Trotting Day! Races uneventful except I lost two pounds on my last two horses— one bolted, one tripped. Lordship won the Cup. Caught the evening train to Lyttelton in high spirits* [after imbibing spirits at the track?]. *Uneventful ferry passage. Wellington fairly warm and windy. Train to Upper Hutt valley, three shillings. Large variety of rides up to Napier that day. Many racing horses in the pastures. Stayed at nice beach* *house in Napier out on the coast, a very civic-minded town, plugging for tourists. Disco coffee bar that night filled with Norse and German merchantmen, the Norsemen extremely drunk, rowdy, and generally comical. A few leering Kiwis, including three overweight homely girls, and one cocky Aussie boy—real tall with curly hair. Amusing scene. All the Norse boys had bottles hidden under the table or elsewhere on the premises and one by one the proprietor would locate them— a game lasting all evening as the boys were sneaking to get the others.*

NOVEMBER 10. *On to Rotorua via Lake Taupo—three rides make it. Nice, clear, warm afternoon. Arrived Rotorua 4 p.m. just in time to visit the hot springs and geysers and the Maori village. Rotorua dead at night. Smartly went to see The Americanization of Emily [starring Julie Andrews], very good. Fooled around after and gave up.*

NOVEMBER 11. *On to Auckland, easy hitchhiking. A long ride with a fisherman who had some freshly smoked rainbows he had caught near Taupo. Incidentally, at Rotorua I went to see National Kiwi Hatchery at Rainbow Springs where some of the perfectly clear springs held big trout—browns and rainbows more than twenty pounds! Magnificent view through underwater window. They fed some of them feed with liver in it. At eighteen months, they could reach eight pounds in weight.*

NOVEMBER 12. *Rick and Sally back. Went to party with bashful bus driver Roger, a young suburbanites party, not a very attractive bunch but weirdly interesting (takes all kinds!). Didn't pay much attention to us as we sort of crashed the party immediately after the barbecue. Oh, well. Got severely tight guzzling all the beer to be found in the* *kitchen. Singing folk songs to the accompaniment of a well-played ukulele, 90 percent American tunes. Roger has tiny flatbed truck and grows mushrooms.*

NOVEMBER 13. *Awoke at 6:55, dashed my clothes together, paid up and left—my taxi had been waiting for me at the steps for the past ten minutes this Sunday morning. Thank God I didn't have to buy an out ticket from New Zealand. Barely squeezed by, thanks to the Air New Zealand ticket counter manned by a Yankee. Bus trip fare to the airport, $1.20 fare—outrageous.*

I found New Zealanders very friendly on an acquaintance basis but more reserved longer term—very English in dress, attitudes, morality yet they disparage the Pommies. New Zealand has a big labor shortage, making services very inefficient or even absent in places. Self-service for many things. High prices except for dairy products. The country is conservative and about twenty years behind in modernity. Young people seem square and unknowing, brought up in Christian Puritanism and socialist middle-class morality. They defend this as the only way accepted—dressing a la Brittania '66 in miniskirts, mod clothes, long hair.

FAIR DINKUM DOWN UNDER

SYDNEY, AUSTRALIA. *Arrived about 11 a.m. Sydney time. Checked in to People's Palace, a plain dingy place. Great name. Sunday in the city is dead, most places closed, so I wandered about. The people seemed very friendly except that it was a city built by White Anglo-Saxon Protestant (WASP) immigrants but nearly devoid of Asians or any other race except for Aboriginals.*

NOVEMBER 14–18, 1966. *Looking around. Enthralled by good-looking busty girls everywhere downtown. Many blonds. Went out to Bondi Beach once and Manly Beach once—windy days. Nights at Kings Cross or the beer lounges downtown. Days checking on jobs, no luck so far. I had decided to stay and work in Australia, I have no memory why. But the available jobs weren't my sort.*

NOVEMBER 19. *Saturday. Flat hunting—finally found one at Kings Cross, pretty well satisfied, $7.50 per week. My first rented—not assigned, borrowed, shared, or overnight space since living at home. It was all mine, but in hindsight a pitifully small, ugly thing down a dark street. I had to put an Aussie one-pence coin in a slot above the cast iron steam radiator for it to cut on and heat the room a few minutes. I only hung around there to get shut-eye, and of course would either wake up cold or wake to put in another pence and nod off again.*

Went downtown and heard some very good Dixieland, and the beer was only twenty-two cents a glass. Met a fellow who gave me a ride back home and treated me to a beer and sandwich, but not a very interesting sort. The Aussies were very friendly to Yanks and eager to talk to me, as we were the antithesis of colonial Britishers and had "saved them from the Japanese." Sydney in 1966 had not yet been designated an R&R military leave destination so young Americans were still a curiosity in town. That was all to change shortly when troops on their two weeks R&R leave from the war began pouring in to town.

NOVEMBER 20. *Sunday. Bondi Beach, beautiful clear blue day, thousands of bikinis. Talked to a thin blond for an hour whose father was a welder in the country. All about American slavery: the economic system/intellectual rigidity/theological dogma/legal repression/class convention/materialism. And don't forget social timidity and hypocrisy and affectation/complacency/pomposity. Slavery notes must have come from something she was reading.*

Cheerful to walk through Kings Cross, immigrants and shops of all types, crowds of people, a festive, holiday atmosphere with my little room hidden in a corner street. But why all the excitement? Are the people of Kings Cross bored, aimless wanderers? I hope not for future's sake. [Here, I may be unaware that I'm reflecting on myself rather than the people of Kings Cross.]

NOVEMBER 21. *Called for job interviews, got library books. I feel that since Vietnam I have walked and wandered many miles, and bought many beers and cups of coffee unconsciously searching for companionship. I have often been lonely, even often in college, like many other students in school away from home. But what about the*

*girlfriends? I suppose the months without girlfriends must
have far exceeded those times. The difference now is that
I am a much more mature and stable (and battle-tested!)
person.*

NOVEMBER 25. *Last night was Thanksgiving, a dead but
pleasant holiday. Two days ago I "sprained" my thumb at
Bondi Beach while body surfing—violent pain for about five
minutes. I walked out of the surf, my eyes fogged, almost
passed out, didn't know if it was the shock. Pain eased off
after a while but then the swelling started—still swollen.* [I
never went to a doctor but used an ice pack. I still have
a bump behind my right thumb where it probably broke
and healed off-center.] *Today cashed my $850 check, quite
an operation.*

*Impressions—terribly aggressive flies in Sydney everywhere,
and when you swat at them they head for cheeks, eyes, and
lips seeking moisture. Tonight is Friday night: Every good
Aussie gets reeling, commode-hugging, shit-faced. A nice
mess when the bars close down depending on the week day,
with strict dram-shop laws. Guys go in and out one door,
females (birds) use a separate entry and stay segregated
by a wall through the middle of the long bar, probably the
source of surface hostility and teasing by the two sexes.*

*New Zealanders are much more English, Aussies extremely
American in dress, speech, movies, music, attitudes,
everything. Fewer London types here—miniskirt, pancake
makeup, peroxide, and accessories. Coffee delicious here,
especially the cappuccino. And the women definitely have
bigger busts. Everyone gambles here, on the horse races, via
off-course TOTAL betting, the opera lottery, the national
lotteries. Devil may care.*

DECEMBER 8. *It's the anniversary of Pearl Harbor on this side of the date line. Took the job at Grace.*

DECEMBER 17. *I finished my first week's work at Grace. My diary's pages skip lots of days in my first month in Sydney, with most entries being quotes from writers like Thomas Wolfe, Gustave Flaubert, and T. S. Eliot as I did a lot of reading while searching for a job and finally succeeding. I had started working in the research and development department at Grace, the largest retailer in Australia, which was modernizing its business model from old-fashioned hand ledgers and bean-counting to track inventory, accounting, and such. We went up to inspect their warehouses and were astounded to find of dusty stacks of decades-old sales items off the company's books.* The TV show American Pickers would have loved it!

I made friends with two Aussies in the office who were fun and hilarious with down-under wits. Went "brim" fishing up at Hawkesbury River last night with Tex and Martin. Caught only catfish. I never found out what species of brim was in Australia. I had put my new friends up to an exploratory fishing expedition and they were game but certainly weren't fishermen.

Went to see Rick and Sally this morning, unfortunately in their defensive mood jabbering about the unpopularity of Americans around the world. [This comment is strange in that traveling Brits at that moment in history were often not well received—on the other hand, the unpopular Vietnam War was on and they were peaceniks and I was a veteran.] *I am very defensive about America around them because I instinctively recognize their criticism is malicious, not objective, the product of their worrying about or their unconscious dislike of their home country (fleeing their homeland?). However, my own attitude is increasingly critical. I am afraid America is overbearing in her influence—economic, political, cultural—and one day many nations may be willing to align with the oppressor communism to avoid the assault of mass culture, to evade industrial and financial domination.*

[More pontificating!] *I often think the foreign policy that would best benefit all nations and America in the future would be this: discontinue all military and political interests except those necessary to protect the physical US, and devote those freed funds entirely to humanitarian and cultural activities. Health support, anti-starvation, medicine, birth control, sanitation, education, and direct investment in housing, communications, transportation, and machinery. This doctrine would assume the first concern is the physical welfare of people. Communist influence is now on the wane. Nations would turn to the West to bootstrap them into modern life. US citizens would have to accept some sacrifice in their growth in standard of living.*

In Kings Cross this evening, I saw a nice-looking man in his forties in shabby dress on the curb, wolfing down a bowl of spaghetti, beside him an old Army blanket roll and a discarded leather golf bag for a suitcase. It struck me as a pitiful site and I think if he had begged I would have given him something. It's strange that he moved me more than the common sight of Asians en masse in abject poverty! (There but for the grace of God go I?)

"In literature, ideas leave their cloisters and descend into the dust and heat to prove their virtue anew. [Source: *Understanding Fiction* by Cleanth Brooks]*"*

Vietnam has estranged—warped?—my personality and changed me to a solitary sort. The aloneness, however, I cannot continue. I don't know what I am trying to spit out here, just self-analysis coming from wandering around the world mostly alone after the weirdness of a full year in Vietnam, the inevitable introspection of a lone traveler, life in transition, moving through strange places.

DECEMBER 19. *I met Ltjg. Mike Colmes, on thirty days leave from Saigon, as he was walking through Kings Cross last night—what a coincidence! What is the story on Harvard professor F. O. Matthiessen killing himself, followed by Cesare Pavese?* [Matthiessen and Pavese were two literary luminaries who died by suicide in 1950.]

DECEMBER 20. *Three major American illusions inducing travel: (1) the very American desire to escape America and fellow Americans, (2) the American tendency to confuse some particular place with an imagined Utopia* [e.g., Australia—at this point I was likely disappointed in the sordid, touristy Kings Cross district of Sydney]*; and (3) the perennial hunger for absolute freedom.*

My blue diary entries for Australia end there, and the next entry is four months later when I am on the island of Bali. I recall that Christmas, however, when I dropped by Rick and Sally's pitiful hole-in-wall flat to share a so-so holiday plum pudding at their invitation. I suspected that they hadn't gotten far meeting new people so I remained the only handy acquaintance around. Indeed, they seemed lonely, remorseful, and homesick—I suspected that they didn't much like Australia.

I assumed I would never see them again; I had gotten work, made friends, and stayed in Sydney well into 1967. An entry in March noted that I went to a cocktail party given by them but I have no recollection of the party and I know Rick and Sally had neither money, place, nor inclination and friends to throw a party. Toward the end of that year, word got around that they had returned to their hometown in England, which turned out wrong.

I was hanging out a lot with my Brit friend Mitch who went to work with me at Grace about the same time. Mitch had a little Morris Mini and we zoomed around the cliff roads of Sydney at night, usually pretty potted and probably more hazardous than the average day in Vietnam, but he never got a flinch out of me, which made his driving even more deranged. He eventually ended up dating my girlfriend's roommate.

I recall that we went to a fancy party where I wore my new green (!) tropical weight wash-and-wear suit bought a few months earlier in Hong Kong, discovering it crumpled next day on the floor with a huge red wine stain running all the way down the jacket and down the pants leg. End of backpacking suit.

I had gorged on some pretty exotic food in Hong Kong, but Sydney Harbour had its points. Several seafood restaurants, with great local white wines and local seafood dishes, overlooked various beaches along the cliffs overlooking the harbor. Lots of ambiance outside on a full moon. Ozone Café on Watsons Bay. I was still in crazy about the small Australian rock oysters first served to me in Hong Kong. Snails, steamed mussels, bouillabaisse, lobster and crab paella. They plated up some delicious local fish filets that I have never since eaten. I would look for the same fish on the menu at the next restaurants I visited.

Karen and I went down to Bondi Beach at 9 p.m. and ran barefoot along the water's edge for exercise. Ate supper along the cliff in an Indonesian restaurant—almost empty. I was disappointed with the chilly breeze, the hot busy summer having surrendered to fall so quickly. In my last few weeks there, I was looking at the half-empty beaches and forlornly recalling earlier times.

FEBRUARY 5, 1967. *Sunday. Went down to Canberra* [Australia's capital, south of Sydney a few hours and more temperate in climate with mountains nearby]. *Trout fished in Murrumbidgee River. Went to the Australian Gran Prix with my friend's degenerate friends, mostly Brits.*

APRIL 2. [I was preparing to leave Sydney.] *To do: cash check, shots, write aunt, mail clothes, buy gear, oyster party, library books, sell camping and fishing gear, buy radio, spearfish club, water tablets, pear soap, ear plugs. Aussie sayings: Keep the faith, baby; fair dinkum; beaut; good on yer, mate; you'll be right; drag queens; butch girls.*

APRIL 8. *Saturday. Departed Kingsford Smith Airport for Singapore at noon on Qantas Airlines after two hurried brandies with Karen, Mitch, James, and Pat seeing me off. Very sorry to leave; however, excited about my future travels. Sydney was wonderful especially the last two months, but I feel I was there too long really. I flew out with a new compact 4x5x1-inch diary in my backpack. I still had my 1967 onionskin pocket notebook but must have been thinking ahead that I would need bigger paper to take the more extensive notes on my travels.*

My pocket notebook continued to record thoughts on the hoof, acquaintance names, phone numbers, and quotes from my reading. Such as "Like most women I don't believe in ideals . . ." and "The fullness or emptiness of life will be measured by . . . ." Why? For what reason was I jotting down all the supposedly memorable words I happened across?

For most of the next 10,500 miles from Sydney to Paris (the distance as the crow flies), I traveled mostly in shorts and military store plastic flip-flops and shouldered my small backpack. Half a century later, I had two total knee replacements. Which worried my knees more—running marathons or touring the planet in three-dollar footwear?

BACKPACKING TO EUROPE

ARRIVING IN SINGAPORE, I was taken aback by suddenly being in the midst of a surprisingly modern city, built western style but resolutely Asian. In the middle of town, I walked along in stifling heat with my new backpack until I stopped at a street stand for an iced lychee nut drink—just about the most delicious, thirst-quenching concoction I ever had. Or I was just very dehydrated! I soon found a cheap night flight on Thai Airways to Java, then flew on to Djakarta (now known as Jakarta).

> **SUNDAY.** *Djakarta. The city is quiet and empty. It sits on a flat field with a few imposing monuments and modern buildings standing quietly, spaced by grass and shacks and broad empty avenues here and there. Some unfinished structures are already deteriorating, some streets run nowhere. It is hot and dry this time of year. Walked the streets and went for a swim at the hotel later, many Yanks poolside, some Japanese. Moved to a smaller hotel the next day—only 500 rupiahs.*

> **MONDAY.** *In Djakarta. Got visa extended. Truck ride back from Chinatown with Dutch missionaries. Reading Bruce Grant's Indonesia. Mailed letters, people rather friendly.*

> **TUESDAY.** *Departed Djakarta for Bandung, traveling by truck. Heavy traffic on roads, many potholes, some repaired in places, abandoned vehicles, good numbers of*

*soldiers. Rice, bananas, pineapples, smell of cloves. Many
recent model cars, everyone well- or decently dressed, wood
or concrete block houses with glass windows, and some
brightly painted bungalows. Bikes, cycles everywhere.
Resort areas here and there with pools, restaurants, etc.*

*At Bandung, my hotel Panghegar comfortable, tastefully
furnished Japanese-style with an inner courtyard, stone
walks, orchids, delicious meal $1.30, TV set, very nice oil
paintings, English-speaking student at desk, I left him my
address.* [I don't know why I underlined the last part.]

WEDNESDAY. *Left Bandung at 1030 after having been
giving wrong time for the early train of 0630 when it was
already gone at 0600. New coaches and Krupp diesels;
there are still some Dutch steam engines on the rails, too.
The woven straw back seats are fairly comfortable. Nine
hours ride to Jogjakarta. Bought two rice rolls and five
pieces of peanut brittle for ten rupiah. Passed mountains
and a couple of volcanoes, no smoke visible. Rice, bananas,
palm, papayas, and avocados on sale. Hotel Garuda
shoddy, no water in room—suspicious house boys. Visited
Nitour, expensive, vacant "official" hotel. US$20 cash gets
2,100 rupiah at the bank. Took public bus to Borobudur.
Cigarettes here have clove mixed in with the tobacco, and
Java in general often smells like a wet wine cork—know not
what it is.*

*Borobudur itself is impressive, over 500 man-size Buddhas
in various niches and "bell-cages." Perhaps a million relief
figures in river stone. Four Westerners at public exhibit; the
only visitors despite the cheap admission. The guest book
had many Germans, some Americans, Russians, Czechs,
and others. Ate a red papaya by the side of the road while
waiting for the bus. A boy offered me his sister—she was*

flirting. On the return bus, met a Chinese mechanic who had lived in Brooklyn during the Depression, home is Kuala Lumpur, has lived in Java for five years, robbed in Naples once, bounced around India, Honduras, and elsewhere. We shook hands and parted.

Beautiful big orchids growing everywhere. Hotel Sarkies. I took an easy, fast train trip—six hours to Surabaya then got a 75-ringgit [Malaysian money] *ride in a betjak to find a hotel. Lousy lunch, slept in afternoon. Bought a microbus ticket to Bali for 450 ringgit. Ran into Rick and Sally (!)—this time off the boat from Port Timor. Nick looked, and said, he felt feverish, he was unshaven. Rick and Sally sounded inclined to take the same bus for Bali. Staying at another hotel. I drank beer in a restaurant named Sputnik, guitar and rattles players serenading Latin-style music.*

SATURDAY. *Departed 6 a.m. on the microbus sitting next to Mrs. Leo R.—mid-fifties, married but very much a world traveler, very chatty but interesting thank God, and sensible. No other Westerners. One very cute Chinese girl about twenty-two—hostess type. Long trip along Java coast, roads medium, vehicle good. Talked, talked, talked about travels, countries, morals, and such. Arrived Bali Hotel after dark at 7 p.m. Supper, then Gamelan music and Balinese dancing for a couple of hours. Gamelan is the traditional ensemble, mostly percussion music, of Java. Dancing is neat, much more interpretive with more fluid motions than other Asian dances.*

SUNDAY. *Swam in the morning at Sanur Beach. This place is void of anyone but locals, beaches, hotels; streets quiet and often deserted.* My diary doesn't elaborate on the circumstances making Indonesian cities half deserted. I arrived just after the coup by General Suharto over

President Sukarno, the founder of modern independent Indonesia after Japanese collaboration in WWII and breakaway from Dutch colonization after the war. However, Sukarno gradually took his country left and fell under Communist influence from a benevolent dictatorship, resulting in a coup by his generals with Suharto taking control and throwing Sukarno in prison.

The country had been through a bloodbath purging Sukarno loyalists and randomly killing and imprisoning many foreigners. After I arrived in Bali, an acquaintance warned me that I was one of the few non-natives walking around; most foreigners had gone to police stations and put themselves in jail to save their skins!

Indonesian lunch very good at a local cafe. In the afternoon, went to a cockfight, and later watched a Hindu procession with boys in trances with knives.

MONDAY. *Spent morning looking at art galleries. Moved my stuff* [Stuff? One medium backpack!] *out to Sindhu Beach Hotel and had a swim. In the afternoon, I roamed around the town, Denpasar. In the evening, I talked and dined with two US women, one a Peace Corps teacher and the other a missionary teacher of English. I met Jack, an ex-Marine officer and Nam vet who had spent two years traveling through the South Pacific islands by yacht. He had worked in Tasmania and Queensland. Jack was a taciturn, low-key, athletic-looking guy who seemed more homebody than world wanderer. He told me much more after I got to know him. After his tour in Vietnam, he met two fellow Marines in Honolulu and together they scraped up funds to buy a fifty-five-foot schooner headed for the auction block. It seems a California family had bought it new and sailed for Hawaii but ran into a gale. Getting de-masted and* *nearly sinking, they managed to get towed to Honolulu. Their maiden voyage killed their love of sailing, and the damaged boat was put up for sale. Jack and his buddies refitted it, and set out through the Hawaiian chain. For the next two years they hopped islands through the South Pacific, and sold the boat on arrival in Sydney.*

TUESDAY 6:30 A.M. *Departed in a green DeSoto for central Bali with Jack and a Mrs. Ripper. We visited a woodcarving town, the rebuilding of a temple next to springs, a hermit cave, a volcano, and a wooded lake, then went on to Ubud to visit an arts center. An American painter named Blanco showed us around his open villa home, and we met his wife, a Balinese dancer, and two children. I bought an elephant statue and an ebony hand.*

WEDNESDAY. *Met Jack at the Bali Hotel in Denpasar at 8:30 a.m. Our guide lent me his bike and Jack rented another one for 100 ringgit.* [After the coup, Bali was almost devoid of tourists.] *Cycled to Kuta Beach, about thirty kilometers south. Nice surf, no one anywhere. Very hot day, bike so-so, very tired when we finally got back. Had beer and lunch, checked some woodcarving shops, and sat and talked about Europe, Vietnam, and more while drinking. I caught a Bali Beach Hotel bus back to within five-minutes' walk to Sindhu Beach. Many mosquitoes at the beach. Supper at the hotel with Mrs. Ripper. Swiss honeymooners joined for another beer. Returned the bike, talked to the guide's English class, and he thanked me and US family here plus one Scandinavian-looking fellow.*

THURSDAY. *Mrs. Ripper and I met Jack and the guide at Bali Hotel and went by public bus to watch a Barong dance—this one a Hindu mythological play at a temple eight kilometers north of town. Very good acting and costumes,* *beautiful temple front for background prop. With the show over at 11 a.m., we continued on twenty or thirty kilometers to a small town, Klungkung, had a Chinese lunch and returned to Denpasar. Later grabbed a Fiat out to Sindhu Beach, packed, and returned to town walking a mile with my pack. Watched Legong dances in the suburbs at 9 p.m. with Bill and Jack, and caught a bus for Surabaya at 3 a.m.*

The bus to Surabaya was twelve hours long on horrible roads. Difficult to find a cheap hotel and finally stayed at a Chinese losmen [the lowest, unrated category of hotel in Indonesian government ranking—similar to a hostel]. *Supper at the Chinese restaurant was very good. Washed clothes—no hot water for ten days now. The many students were very helpful in Surabaya and they reminded me of US citizens for some reason.*

SATURDAY. *In Surabaya. Bought train tickets with Jack. Train first class—750 rupiah to Djakarta leaving at 4:30 a.m. Surabaya an unenticing place except for the great Chinese meal last night. Sunday arrived Djakarta at 7:30 a.m. and went on foot and by becak* [Indonesian rickshaw] *to Hotel Indonesia to buy our tickets to Singapore. Later visited a fairly good museum showcasing Indonesian artifacts, Chinese porcelain, and native New Guinea and Borneo items.*

SUNDAY–WEDNESDAY. *Back in Singapore, wandering around midday, eating delicious Chinese food in small open-air cafes. The ripe lichee nut juice on crushed ice sold in street stands is hands-down the best thirst quencher I have ever tasted. Just about keeled over with the first sips. Small world, saw Mrs. Ripper briefly at the post office. Heavy rain in afternoons. Went to the Great World Fair* *then hung out at the Golden Venus bar at Hotel Orchard with Jack. Leaving for Kuala Lumpur tomorrow.*

THURSDAY. *Departed by bus across the border to Johor, Malaysia—seventeen miles. Caught a one hundred-mile ride after thirty minutes walking and waiting with a service Land Rover, then a twelve-mile ride, then a ride with an Indian palm oil plantation overseer in a VW for seventy-five miles. Then a ride with a (protestant) Asian missionary into Kuala Lumpur. Missionary had been in China in 1949–1951 and was expelled (survived!) with the other missionaries. Arrived at 5 p.m., walked around for two hours in the evening.*

FRIDAY. *Went shopping for bush jackets, then visited the tourist bureau. In the afternoon, we took a $2.50 bus tour—pewter factory, Batu caves (three caves in a limestone hill outside Kuala Lumpur), a tin mine, a rubber plantation, and the university grounds. Kuala Lumpur is an attractive city, spaciously laid out with modern buildings everywhere. Many new cars and bikes on good roads, new petrol stations—Esso, Mobil, Shell—busy, full shops. Supper at a Chinese restaurant then saw movie Georgy Girl with Lynn Redgrave—very funny, good.*

SATURDAY. *Departed, walking twenty-five minutes across town to the bus station and took an eighty-cent bus ride twelve miles out to Rawang. In fifteen minutes, caught a ride with a non-English-speaking Malay in an Austin all the way to Penang at sixty to seventy miles per hour. He bought us nasi goreng, a fried rice dish, at a Chinese rice shop. This is a beautiful interior city, very suburban and middle class. Lovely, lush mountain country for the fifty miles before arriving in Penang, lushest jungle I've ever seen except Vietnam Highlands and maybe Panama railway across the* *isthmus. Rubber trees beautiful in rows, dark, dark shade underneath, splotched with sunspots. Arrived Penang at 3:30 p.m. on a car ferry. Terribly hot. Found a Hotel Noble in Market Lane after wandering about an hour, one dollar per day per person. That night we ran into a snack shop serving great hamburgers in a US atmosphere.*

Penang is a quiet, lonely city. Many colonial buildings, broad streets, parks, nice hotels, and two or three cocktail lounges. In the morning, we took a bus up to the coast for a swim and looked around, then came back at 4:30 p.m. Supped again at the snack shop and headed to a bar for beer.

MONDAY. *Looked around town. Met a Peace Corps boy from Thailand who gave us some travel tips. So we bought $6(Malaysian) tickets up to the border and into Bangkok. Penang is a lovely city, quiet, modern, a good place for honeymooners. Bought Letting Go by Philip Roth.*

TUESDAY. *Jumped out of bed at 6 a.m. at Hotel Noble— nobody woke us, train ferry leaving at 6:30 a.m. Quickly dressed, tossed out packs full, and ran through the streets and alleys—made the ferry with one minute to spare! Pretty morning, quiet briny breeze, only a few people on the ferry. Large Chinese sailing junks in the distance and small power junks ferrying people across the straits. Pods of bait fish here and there.*

We rode in a small, quick, noisy diesel train to the border at Padang Basar arriving at 10:30 a.m. and then sat and waited for the Thai train until 12:30 p.m., boarded the packed full train, and rolled very slowly up to Hat Yai at 3 p.m. Debarked, bought 6 p.m. tickets for Bangkok, and had a very good lunch—a nasi goreng-type dish with loud western background music. Hat Jai a sleepy town, 6 p.m.

*train also packed full. I finally found a seat, Jack in the next
car. Mostly read Roth, slept about three hours in the early
morn.*

WEDNESDAY. *Still going on the train, very tiring on
the lower spine.* [I'll never forget, and can still feel, the
sensation on my butt made by those woven bamboo
thatch seats. Did I not know what NSAIDs were yet?] *Met
Bill and Nate, two Peace Corps types serving in Thailand
and heading back to Bangkok. Tried their rice wine.
Fantastically hot in the afternoon I recall the train only
went at sprint speed, about fifteen mph, through the jungle,
so little breeze through the open-air cars. My signal memory
of this trip is the unvarying meals of rice with whole Thai
peppers and bits of meat, so hot that we spent the first
minutes picking out peppers with our fingers and tossing
them out the train window. The train trip was so crowded
and uncomfortable.* I recall the journey being three days
and nights, but it was only two.

*We pulled into Bangkok at 5 p.m. and the four of us took a
taxi, Nat speaking good Thai, taking us to a Chinese hotel,
very nice and only US$1.50 each. That night we splurged at
a restaurant with great steaks, martinis, and entertainment,
then on to the Balcony, a US hangout, then to CanCan for a
look.*

THURSDAY. *First full day here: bank, tourist bureaus,
AmEx, lunch, embassy, etc.—all by bus—the city is
sprawling. That night we went out to a beautiful supper
with the Peace Corps fellows at Helena's Restaurant in the
Peninsula Hotel on Surawong Street. Also Lee and Roger.
Roger is Japanese American and past editor of* [University
of Southern California] *USC campus newspaper, also Phil
a very witty Teddy Roosevelt lookalike Peace Corps politics*

*professor at Bangkok University. I ended up with girl at
2:30 in the morning.*

The next day very tired, went to see Funeral in Berlin
*with Nate and Bill—so-so. Slept later than Jack and I had
supper back at Helena's (giant prawns) and spectated at
the Balcony Bar for a while, listening to a very polished
Philippine trio.*

SATURDAY. *Departed Bangkok on third-class train tickets
for Chiang Mai at noon, an eighteen-hour trip. Very hot,
had a good cheeseburger before leaving. Wooden seats okay,
treated to supper by two obnoxious Thai individuals sitting
next to us, drunk on rice whiskey. Arrived Chiang Mai at
6 a.m. and walked with packs out of the railroad station
down the street and two miles across the river in misty
dawn, no idea of destination. Stumbled onto Chiang Mai
Hotel, very cheap and nice, with message at desk for Bill
and Nate there. At 10 a.m. went by bus, a malaria control
team riding along, to a mountaintop Buddhist temple
and the King's summer palace. Hot thunderstorm waiting
to enter the palace gates. Later supper at Pat's Bar &
Restaurant, and shopped in town. Lovely shopkeeper girls.*

MONDAY. *Looking around town, slow place by day. Bill
and I had beers and dinner, and hit the town by becak, out
to a nightclub—nice, very dark inside, about fifty couples
dancing Thai and western style. Later, our betjak took us
on a tour of various after-hours spots, then back to hotel at
midnight.*

TUESDAY. *Caught the noon train back to Bangkok, lucky
to find two seats, reading Fail-Safe then starting on John
Steinbeck's The Wayward Bus. Once again, two drunk
obnoxious Thai individuals in our car. In the evening, Jack
and I spent two hours drinking Mekong rice whiskey, very*

cheap, then sacked until morning. Off train at 5:50 a.m. and stumbled about town with hangovers and packs for two hours looking for a hotel, finally went back to the old Saksewasdi Hotel.

My diary is blank for the next couple of days except for an entry about going to a Thai boxing match one late afternoon.

FRIDAY. *Visited the floating market with Jack, Nate, Bill, and two other Peace Corps guys. Stopped off at the shopping area in the canal—monkeys and bears on exhibit. Later went to visit the large wat* [temple] *and the palace. At noon, Bill came with me to visit my Irish friend Flynn at his office at Rolibec International. Flynn left work and entertained us over beers for two hours, then we all went out to his villa where we played badminton for a while with Abdul from Pakistan joining us.*

Flynn was the same fellow naval officer and true Irish character I met in Saigon; he was the attaché but ached to get out "in the field" like me. Turns out, he got out in the field.

When I was mustering out on my last visit to Saigon, I had to turn in my gear and get a physical. I stayed in Flynn's BOQ room while he was out of town. The room was on the second floor, just over the lobby. While I was out on the town one night, a Vietcong cyclo packed with C-4 explosive drove into the lobby and blew up my room. I gave Flynn my .25 pistol the day I left Saigon. Flynn left the Navy too, and ended up in Bangkok selling life insurance for a day job. However, I later found out he had been making Air America flights into Cambodia/Laos; I suspected he was on the CIA's payroll.

He had had a family back in the States but never told me what became of them.

Years later, checking out of a hotel in Houston, I think I saw Flynn crossing the lobby. For some reason, I decided not to come up behind him, speak his name to see if it was really him, and revisit the old days. I still regret it.

> **MAY 13, 1967.** *If one is not to rebel totally, it is necessary to live within the context of the current social system. Then ten or twenty years ago I had to abide by the laws of segregation, the status quo. Today, my values must change as the system changes. I must be totally for the Black revolution.*
>
> *If this appears hypocritical to you, then I suspect you are a believer in absolute values which places you at the opposite pole of belief from me.* [As a twenty-something, I was following the script that says when you are young, if you are not idealistic you have no heart, but when you are older, if you are not conservative, you have no sense. I like what Clint Eastwood has been alleged to say about reparations: "I never had any slaves, and you never picked cotton, so you're not getting a damn thing from me."]
>
> **MAY 14.** *Rangoon to Calcutta. To get up to India, John and I had to buy tickets on Burma Airways (UBA) and stop in Rangoon—even though Burma was Communist-controlled and American entry was forbidden. Since we didn't have Burma visas, we were technically persona non grata and hostages in sleeping quarters of their choice—the choice happening to be the best choice for some unknown reason. We stayed at the historic British colonial Strand Hotel, which I recall being the background for Graham Greene and Somerset Maugham novels. We were never permitted to leave the Strand and go out on the street.*

*Departed the Strand in downtown Rangoon at 5 a.m.
by bus. Picked up five other passengers, East Europeans.
Departed for Calcutta at 7:30 a.m. and at 9:30, we lost
an engine approaching land, the co-pilot announcing that
"bad weather necessitates us returning to Rangoon." Crap.
This was not my first choice for being reincarnated as a
Burmese grasshopper. Landed okay, back to the Strand for
a nap, then went to see a huge, ornate Buddhist temple with
hundreds of village shrines encasing individual Buddhas.*

*If you can't do something unique in life, at least you can be
uniquely loved. That is the essence of life for the common
man.*

*The Strand is well kept up but plainly furnished, with
mosquitoes in the room and nets for each bed. The place
was nearly empty, and joining us in the cavernous dining
hall was only one other table with four Russians. Each table
had four hovering waiters. After dinner, we went up to the
rooftop for firefight entertainment with tracers crossing the
outskirts of the city as Chinese Communist forces pushed
Burmese troops back into the city. We have decided not to
hang around here any longer than necessary.*

*America as a country needs a constant atmosphere of
inspiration, from a man or a situation. At this time, it has
neither.*

MONDAY. *Departed again for Calcutta—about six
people on train. Arrived with stomach pains at 10:30
a.m. and waited for UBA bus in the airport lobby for an
hour.* [It's unclear whether we got to India by train or by
air; I can't remember which. As for the stomach pains,
this is the only instance I can recall of being sick since
leaving Vietnam. I was eating a wide variety of food
from Indigenous restaurants, homes, and off the streets,

but I carried a big stock of one-a-day tetracycline pills saved from my year in combat, which I think everyone in country was issued for the duration of their tour.] *The customs people were nasty, arrogant—bad first impression of India. We walked to the Salvation Army with Tony, a very short and slow Englishman. The rooms were okay.*

In the afternoon, we got a visa in under one hour for Nepal, visited the zoo, and picked up the mail—two from Karen, two from Dad. The zoo animals look better fed than the locals. In the evening we had a good, cheap forty-cent Chinese meal. The next day we went to buy railway tickets and saw Shirley MacLaine in a spy movie. In the evening, another Chinese meal with a German fellow waiting in Calcutta for more money.

WEDNESDAY. *In the morning, we visited the Victoria Memorial, quite impressive architecturally with fine paintings and exhibits of the age inside. Many prints and paintings. Picked up train tickets, wired for money. In the evening we saw a horrible movie, Rampage, about trapping tigers in Malay—intrigue, etc. Come evening, we caught the third-class sleeper at 10:25 p.m. People everywhere on roads, pitiful sights of life. I distinctly recall early morning backpacking to the train station and counting seven dead bodies—stepped over them on the sidewalks. I sidestepped lepers and one-handed or one-armed or one-footed beggars. I said to myself, I'm going to count the dead bodies I step over and remember the total. Seven. Sad, sad.*

Calcutta to Raxual, India, on the border. Had a derelict arguing with everyone in the third-class sleeper giving us problems. Yankee ex-soldiers in backpacks are captive targets on English-speaking Indian trains. At noon, we changed to a day train. Bob, an Englishman, was traveling with us in the last third-class coach, packed with humanity, people on the roof and hanging out windows, no one with tickets or checking tickets. No safe water, everyone drinks steaming hot tea from freshly made clay cups from vendors who bake and brew at every siding. Then you toss the cups out the windows smashing them in the rocks.

Extremely hot going through North Bihar. I don't know if these particular first-class cars have A/C, but we aren't going to pay the higher fare. A bad drought, thick clouds of dust, no grass, starving cattle and people. We were tired of tea and low on water in our canteen, dirty and thirsty. Changed at 6 p.m. for a one-hour ride into Raxual. A band played in our coach. We had supper and bedded in the railway station.

Raxual to Kathmandu. Up at 4:30 a.m., Bob, Jack, and I depart the railway station by horsecars with our packs and Bob's guitar working our way through six or seven customs and immigration checkpoints. (People are escaping to here?) Waited for a bus until 8:30. A mad Sikh steering wildly, and we sat up front for twenty rupees each. A sheer drop-off from the road, to 8,000 feet at one place. All day by bus, paddies are dry with wilting wheat crop. Gurkhas along the road, many look like Tibetans.

Arrived Kathmandu at 6 p.m. Found THE CAMP (!)—a mean little concrete room, one bed and two straw mats. Next day, we bought local money and walked about town. We went to the blue Tibetan Cafe at 9 p.m. Smoked some pot [my first hashish] *with Tony and a girl and boy who were very high and all giggles. Tony was slurring his words and unresponsive until one long drag, then keeled over laughing. Jack and I continued smoking, no effect. Retired to room, brushed teeth, then it hit us: lost sense of direction in dark, time sequence mixed, rationality cycle, loss of train of thought, rambling ideas, fight to retain rational control. The first sober moment, I found myself inexplicably standing outside next morning at a public water fountain brushing my teeth.*

SUNDAY. *We bought Air India tickets to Patna then caught the noon bus to Bagdeon, got off the bus and walked nine miles up to Nargakot with only two three-minute breaks in three-and-a-half hours climbing time. I'm stronger in the legs than I figured but pulse stayed at 110 for a long time after reaching summit. Cold showered. Dinner with Sherpas and two English Peace Corps boys and one hairy-legged Cockney girl. Bed at 7:30 p.m. and up at 4:30 a.m. to see sunrise in the Himalayas above the Kathmandu Valley. For a while, it was too misty, but then it lifted to show the entire snow-capped chain except for the peak of Everest itself concealed behind a cloud.*

A beautiful view straight down 3,000 feet to green paddies, Sherpa trails, goats everywhere. Sherpas carrying huge bundles of wood, women with rings in noses. Read in the sun in the mornings. After lunch, waited out thundershowers and took a truck down to the base.

TUESDAY. *May 23, 1967. Buddha's birthday. Rented bicycles and cruised around town. In the afternoon, we went up to a temple pilgrimage on the edge of town. The people we met staying at the camp were Japanese, Americans, English, Danes, and Germans. Ravi was the proprietor.*

THURSDAY. *We departed Kathmandu at 1 p.m. with the dust blowing throughout the valley. Arrived an hour later.* [Obviously we had a direct flight; today, a traveler must make the 200-mile trip via New Delhi or spend six hours or more on the road or train.] *We spent the afternoon walking through the HOT, DUSTY, DRY, IMPOVERISHED capital city of the Cooch Behar State looking for a moneychanger at black-market rates, to no avail. Never felt so dehydrated. We had an argument getting into the retiring room at the railway station. Boarded the 9 p.m. for Benares.*

We arrived in Benares at 4 a.m. and slept for one hour at the government tourist bungalow for 1 1/2 rupees, then up at 5:30 a.m. for a tour of the Ganges River—bathers at the ghats [riverfront steps to the Ganges], *crematoriums, market temples to gods Shiva and Ganish. Slept the rest of the day.*

We journeyed on by overnight train to Agra, getting some good sleep for a change by paying up for second-class tickets. [I underlined "some good sleep" to emphasize that we had been burning the candle at both ends, especially by low-class train travel. My recollection is that many Indians never even bought tickets, just jumped on and rode the roof whenever it slowed or stopped. They were packed, noisy cars.] *We arrived at the Taj Mahal at 8:30 a.m. and followed that with a tour of the Red Fort adjacent to it. Impressed by the size of the Taj. Built 1631–1653 in honor of* [chief consort] *Mumtaz Mahal, who had fourteen children.* [The lone surviving picture I have of my trip shows me standing in front of the reflecting pool and Taj Mahal in shorts, and my too-small khaki Brit field jacket. Jack took the picture as I had no camera with me—unfortunately, no memorable photos of my worldwide adventures. A year later, after our trip ended and he returned to his home in California, he mailed me a dozen or so pictures from our travels, which eventually got lost. I never heard more from Jack and never was able to locate him again.]

Taj Mahal stop crossing India

We took a hilariously circuitous ride in becaks to multiple train stations, nearly missing our 12:17 p.m. departure. Obviously, the peddlers are mainly for moving bodies in whatever direction seems open. The train was moving and we had to run with packs, no tickets, and jump aboard. We arrived Delhi in late afternoon and went to the YMCA tourist hostel.

The first day we slept till 10 a.m. [Wonder what went on the night before? No mention in diary except at some point we hit the movies with Steve McQueen in *Nevada Smith*.] *The next day was a Sunday and quiet so we didn't do much. We met a forty-five-ish English woman artist at the hostel who had lived in Greece five years. Mostly talked about West Asia. Good cheap beer, bought some lunch, and slept in the afternoon. Delhi is much more cosmopolitan and cleaner than Calcutta. Good cheap restaurants and coffee bars.* [Based on my diary notes, I think our living expenses on the road were likely running $5 a day or less;

the big budget hits coming when we had to fly to get in or
out of a country.]

We obtained three visas in Delhi [my guess: West Pakistan,
Afghanistan, and Iran]. *Afternoons here absolutely too hot
for going out of doors in late May. Air extremely dry so that
lips quickly become parched. Many American aid people
were staying at the YMCA. City is very clean; cheap US
beer at the American embassy.* Jack and I ate hamburgers
with beer there one day—eight beers. Bad hangovers the
next day. Looking back, I don't recall eating any curry
dishes in India. The only explanation must be that in our
travels through the country we were nearly always eating
curried dishes so the seasoning didn't stand out.

Years later, reading the biography *Victoria & Abdul* reminded
me that even though I had journeyed by land across India and
had read extensively, I knew only a smattering about the country and cultures there— so it is for almost everything else, and
so it is for all people. The internet is a handy shortcut for instant
answers (where accurate!) but maybe makes us all shallow
thinkers. Many generalists, fewer experts. Know a little about
a lot of things.

*We left Delhi Territory in early June, headed for the
Pakistan border by overnight train, and arrived in
Ferozepur at 7 a.m. Took a bike seven miles to the border—
one hour for paperwork, three hours waiting for a bus to
fill up and go on to Lahore. Five Japanese students on the
bus, very friendly, but the ride hot, dusty, and uneventful.
Thirsty but had no money yet. At Lahore, no mail for us
at American Express. We finally managed to trade for
some rupees at a Chinese restaurant, and caught a train to
Peshawar after a good meal and shower at the station. Nice
train but the men had to fit in their own car. Arrived at 7*

a.m. after a fitful sleep amid jovial Pakistanis. My uncle
was once stationed in the Air Force there. The Peshawar
countryside is desolate looking, everybody is military.

We reached the Khyber Pass by bus, wicked-looking,
everyone toting rifles and the pass fortified at all points.
This narrow route along the ancient Silk Road trade route
is probably the most famous mountain pass in the world.
Having remembered it from history class or my reading, I
was anxious to see it and Jack was wondering what was the
big deal. It was a strategic gateway into some of the most
forlorn country I ever saw. Along with Genghis Khan and
various invasions came ivory, silk, textiles, and other goods
considered exotic luxuries in earlier times. The Pashtun
clans control the pass and take a toll on all coming through
as a tax for assuring safe passage. One really wants and
needs safe passage—or else.

An indelible memory not in the diary: Everyone was ordered
off the bus for a search. I sat next to an Indian immigrating
on his Commonwealth passport (as did Jack across the aisle),
and most of the passengers seemed to be Indians or Pakistanis
headed for London with their clothing bundles and pillows.
The Afghan guards went right to the pillows and slashed them
open with their knives, hashish stuffing popping out onto the
concrete. Pillows and hashish confiscated, we filed back onto
the bus and moved on. I turned to my seat mate and asked
him, what now, expecting him to be devastated now that his
dowry to start a new life in Merry Ole England was taken away.
Nonplussed, he remarked that it was just his first try. It was
arranged that he would make it through on the third, and was
taking the next bus back to India at Kabul. *It was about fifteen*
miles going through the pass, with snow-capped peaks all around.
Two Pakistanis took us to a hotel of sorts, and helped us change

money. A Kabul family at the hotel very friendly and cosmopolitan. Sorry to miss being able to walk around sightseeing. Had a good bath. Air is cool and clear.

Kabul to Kandahar. Easy trip on bus from 7:30 a.m. to 4:30 p.m. but very cold the first two hours—wonder the altitude here. Snow-capped mountains on all sides, people bundled up everywhere. Ate a sort of shish kebob for lunch, later had rice and bits of this and that at a mud hut stop off. Traveling with a Pakistani engineer on his way to London.

At Kandahar, we walked around town and the bazaar, very few females visible, but girls around five to twelve or so very curious and friendly to us. Nice horse carts here, one bazaar stall had a 1798 rifle and a double-barrel Russian pistol. I have a head cold or sinus problem from the dust everywhere.

EARLY JUNE, *off to Herat, Afghanistan. Not quite so cold this morning. Afghan girls and women never visible but they're out there somewhere. Yesterday's road was American made, today's Russian concrete. Having periodic trouble with the bus. Stopped at a modern, new western-style deserted hotel for lunch—very thirsty. After lunch, three miles down the road we blew a hole in the engine, think a piston and rod shot right through the hood. Passengers pushed the bus for two miles until the futility of five tons of metal creeping through the desert at 1 mph hit home. Waited for the new bus until 7:30 p.m. The young Pakistani passengers are mad at an old man who refuses to sign a petition saying the driver is not responsible for the breakdown. We got to Herat at 1 a.m.*

JUNE 5, 1967. *UAR–Israeli war begins.*

JUNE 8. *Ceasefire.*

JUNE 10. *Letters off to Karen and Mother.* [I must note here my mistaken recollection that I never again communicated with Karen after leaving Australia.]

We are waiting another day in Herat for a different bus to arrive. Many pelts here for sale—snow leopard, lynx, wolf, etc. They want $100 for a good three-meter snow leopard. I was sorely tempted to buy one and pack it out to Europe, but decided that it would likely be confiscated crossing a border. [Years later living in New York, I read it was near the top of the world's endangered species list. Imagine wearing a snow leopard coat down Fifth Avenue.] *We're staying in a clay-mud hovel, a real dump. Quite a few hitchhikers in town. Very cool in evening, what a change from South Asia! Finally we came up with bus tickets after a group of six canceled. Goodbye Afghanistan.*

JUNE 7. *Herat to Mashhad, Iran. Very rough road to the border, arriving at noon. After lunch, many customs formalities, delays, processes, questions, breaks, lines, etc., we finally depart at 7:30 p.m. Eight tablets for cholera! We cross a very windy plain and arrive Mashhad about midnight. Jack and I share a taxi with Lee (US), a Dane, and Klaus (German)—looking everywhere unsuccessfully for a cheap hotel. We finally go to a beautiful train station at 2 a.m. and sleep in the waiting room on benches. Awakened by screaming beggar woman at 7 a.m. We all decided to move on to Tehran, taking a good, European-style train at noon. Lee bought the Dane a ticket, who then skipped. The people in our compartment, young men, having a fight, one is crying. Good dining car. Read The Spy Who Came in from the Cold in the afternoon. In the evening I started on Ted Sorensen's Kennedy. Iranian women interesting, wearing dark capes or western dress. Arrived Tehran in the morning and a multilingual Israeli tourist guide took us to*

*the bus station. We bought tickets for a bus next week to
take us to Ashure, Turkey, via Tabriz, Iran, then found a
cheap hotel and looked around town.*

*Tehran is very modern, full of US cars, wealth here very
visible. Most men in town wear nicely tailored suits with
or without tie, good leather shoes, square toe. Women very
attractive, however tend to big noses and heavy peasant
legs. Many US products—Coke, Pepsi, 7 Up, soaps, etc.
Good local beer, delicatessens everywhere. The traffic is
terrific but the streets are tree-lined and reminiscent of
Paris, much neon advertising, thousands of shops, beautiful
Persian carpets, jewelry, silver work. The language is Farsi.*
[Here I wrote down the numerals one through nine in
Farsi.] *Saw movie* In Like Flint, *very entertaining. Good
sausage and egg sandwiches and we drink soft drinks all the
time.*

JUNE 12. *Tehran. My twenty-seventh birthday. Was
supposed to have called Karen tonight. This afternoon Jack
and I sat in the railway restaurant and drank three beers
each in celebration of my birthday, talking about careers
and such. I went out and bought a French-English short
story parallel-text compendium.*

We moved on to Tabriz by bus. [It was raining, and my
diary reported not much to do, not much of interest.]
*Big argument with clerks over our hotel bill. Good food in
restaurants here. After a couple of days, we decided to move
on to Erzurum, again taking a bus, and then finding a train
going the Ankara, the capital of Turkey.*

*Ankara. Had lunch with Uncle Montrose and his company
cronies.* [Montrose Barrow was my father's oldest
brother, who was building an oil refinery in Turkey at
that time as a senior engineering consultant. I never

knew him well, but he was a reserved intellectual with an artistic and adventurous bent, having gotten on a tanker, gone to Europe, then to California, worked in a movie studio, joined the Coast Guard, shipped out to Alaska, and joined an Arctic expedition. In retirement, he oil painted as a hobby. I later learned that Uncle Montrose also lived at Tanglewood Plantation (like my Dad) for several years, putting up with Uncle Douglas and his "all business" bossing.] *Strolled around town, primitively modern. Turkish women look nice. Good supper with Uncle Montrose and Jack. Beforehand had some beer and conversation in his room. Impressed me as a quiet, happy man; however, he was tired of Ankara.*

On to Istanbul. We bought food to make sandwiches on the train. An easy, uneventful ride. No problems getting around or finding bedding in town. Aya Sophia was closed so we bought tickets to the Blue Mosque, very impressive inside with its blue tile mosaics. Built 350 years ago, it's a standout monument of classical Ottoman architecture, six spires, geometric, gray color, very impressive. [When I saw Muslim and Arab historical sites, I would recall my father's admiration as an engineering graduate student of their scientific and mathematical achievements.] *More so than Saint Sophia. Found cheap places to eat. Shish kebab with bean salad and bottled water—three lira.*

We went over and visited the Istanbul Hilton. I had fond memories of staying at the Hong Kong Hilton and drinking and playing roulette at the San Juan Hilton during the Cuban crisis, so visiting a Hilton in Istanbul was natural after weeks away from familiar American surroundings. We succumbed to a very expensive salad in the snack bar. Turkish men are very inquisitive and brusque. Surprisingly helpful to strangers though. A good black-market here.

TUESDAY, JUNE 20, 1967. *Visited the Topkapi Palace. Extremely good collection of oriental porcelain. Collection from eighth–fifteenth centuries AD. Great fantastic jewelry collection gifts originally to sultans from around the world. Especially encrusted weapons and ornaments of gold with diamonds, rubies, and large emeralds. Looked at the livery stable and the library of old manuscripts. This is the most interesting tourist site here.*

WEDNESDAY. *We took a morning boat up the Bosphorus Straits, a two-and-a-half-hour journey to the headwaters of the Black Sea. Got off at a small fishing village and lunched at a seaside cafe. Beautiful blue water. Returned by bus to Istanbul. Turkey is the only country on two different continents.*

EUROPE BY VW

JUNE 23, 1967. *We left Istanbul, entering Europe by bus. Departed at midnight. About six hours waiting for customs at the Turk-Greek border. A bumpy trip in the back of the bus with poor seats, arriving in Thessaloniki at 7:50 p.m. Took a walk on the waterfront and had supper there.* [I presume toting our backpacks all the while!] *Appears to be a rather modern, European, touristy town with nice shops. No Americans seem to come here.*

We continued on to Athens. Greek record music on the bus incessantly. Stopped at Mount Olympus, river running out of the mountains, chapel, woods, deep gorge. The next afternoon, we arrived in Athens and stayed at a dollar-a-night hotel. The main square where American Express office is located is full of loitering Americans at cafes. Go tour Greece and see American tourists. There seems to be no other sorts of tourists around.

We got a ride down to Sounion on the lower coast, thirty miles from Athens. Got on a very slow bus through olive groves, having intended to take the coastal bus. Got impatient and nervous. Finally made it. Bad day so far. So-so beach at the hotel. We got one-and-a-half hours of sun and lunched late. We toured the Temple of Poseidon on a windswept cliff while waiting for the bus. Beautiful view, many names carved in the stones mostly AD 1850 to 1910—one I found from 1787! Lord Byron carved his name.

Returned by coastal bus. A glum French girl skin diver next to me with a friend.

Athens. We visited the archaeological museum, very interesting with artifacts from ruins and statues from the fourth century BC. On my last day I went up to the Acropolis with Jack and we spent an hour there in the late afternoon. Not too impressed, it was anti-climactic. Most impressed with the women in the tour groups—many French tourists here.

Athens to Patras to Corfu. Spent the night on the ferry to Corfu. [My recollection is that Jack and I, after journeying nearly halfway around the world together, parted company at the port of Piraeus, me boarding the ferry, he heading west through Europe bound for his home in California.]

Got high with Rennie from NYC who works for Newsweek—*or so she says. Seems fishy to me. She's headed to Rome like me. She is poor conversational company (pity* Newsweek!*) and the first American girl I've talked to for any length of time (ten-plus minutes) in two years. Terrible twangy accent grates on my ears.* [I'm surprised at this comment, having endured the harsh Chattanooga southern twang of my dear sweet girlfriend Lauri teaching school near my first duty station in Virginia.]

It started off when I sat down beside her—best choice of a bad lot unfortunately—on the bus. We had supper at a waterfront restaurant in Patras, actually very enjoyable with some nice wine. We arrived in Corfu at 8 a.m. and I said goodbye to Rennie. Roomed for thirty-five drachma with Morris at the Hotel Constantinople right on the waterfront square, beautiful views of the water. Morris is an unemployed Jewish NYC film editor, wonder how good he was at editing. Worked on The Defenders. *He seems to still have some money for travel, hypochondriac, thirty-eight years old, looks very Greek with his mustache. He's not used to being without crowds of people around him.*

Morris and I rented Vespa 50 scooters and went all over the island. Very beautiful caves. Club Mediterranean has resorts in two places. There's one German club here. Monastery on the hill overlooking the sea. A langouste restaurant. We had lunch, chicken and wine, by the sea. Had two major falls, the first one me, then Morris driving, the worst. Morris has a nasty forearm and swollen elbow, his own damn fault! I got dizzy from pain in my knee—the bike fell on top of it. Earlier, we ran out of gas out in the country and did some bike pushing to Pelekas. Two knees, two palms, one elbow. [Fifty years later, I went flying off my pedal bike in Tuscany, Italy. Precisely the same damage: two knees, two palms, same elbow, ten hours in local emergency room with last rites by priest to two people on adjacent litters before getting sewn up at 1:30 a.m. A real crappy stitch job, could have done it better myself, with whisky swigs. Moral: Stay away from Italian hospitals.]

SUNDAY. *Spent the day wandering around town with Morris. When it rained in the morning, we sat drinking coffee in a shop owned by a Greek American, listening to rock, the owner lamenting the disappearance of his US Navy business. Beginning at 7:30 p.m., the cricket field sidewalk cafes start filling up. Corfu seems to be very conservative. We drank some ouzo—very cheap. The food is not too good here, too much olive oil.*

Corfu. Brindisi, Italy. Caught the ferry to Brindisi this morning, still sore in knees and hands. Arrived at 4:30 p.m. and strolled around town until 9 p.m. train. Shared a compartment with two boring southern schoolteachers and a bottle of wine. Arrived in Rome, found an okay pension and slept until noon. Then went down to the American Express office.

Rome. Foro Romana. Colosseo. Walked around town. Via Veneto. Spent most of the next day at Piazza San Pietro, the Cathedral, the Sistine Chapel, and museums. [Having thoroughly toured the Sistine Chapel in a new century on a custom tour, I was shocked to read that I had been there over fifty years ago—so much for indelible memories!] *Sold Eastern and Xerox stock, afraid of future news on business and economy.*

FRIDAY. *July 7, 1967. Met Lucia at 9 p.m. with Rosetta. Went to Trevi Fountain, the Campidoglio, then dropped off Rosetta and drove out to the coast for supper. Very pleasant drive; she drove out I drove back. Got in at 1:30 a.m. We had scampi for supper with garlic bread appetizers and nice rosé. Café overlooking the sea. The next day Lucia picked me up at 4:30 p.m. and we drove into the hills in Castelleria Romanas. Several high villages, two beautiful crater lakes. Then drove down through the hills and out to the sea to visit a castle right on the beach. Then back to Roma, pizzeria, and movie Grand Prix in Italian.*

SUNDAY. *Got up late (came in at 2 a.m.) and washed clothes. She's taking parents out today. Will call her tomorrow after work. Studied my French. Lots of Navy, US and Italian, sailors in town. Otherwise, quiet except for tourists. Wish I had a car so I could get out a bit.* [Wait for what happens when I depart Italy for my next country.] *Went to Via Veneto in my gray summer suit tonight, first time I've worn it. Not impressed by the people or fashions on* *the Via Veneto. Probably more chic in autumn after gawkers go home.*

MONDAY. *Left Lucia a note at Silhouette, no use staying in Rome. Sorry that we didn't click—actually we did except we needed something to break the ice. My Aunt El Barrow, Uncle Montrose's wife, who was then living in Rome but on summer vacation elsewhere, had gotten me the blind date with Lucia when Uncle Montrose let her know I was passing through Rome.* [My recollection of Lucia is that she was pleasant and nice looking but somehow I wasn't attracted to her.] *Just a* nice buddy to squire around her town. [Never heard more and my reading the diary a half century later shocked me that we had actually spent more than just one afternoon lunch together eating scampi.]

Saw an Italian boy outside American Express office get down on his knees mimicking worship in front of two sexy American blonds walking by. They kept walking and he got up and grabbed one very roughly by the arm. Tempted to pop him, my better judgment propelled me down the street. Imagined local headline: "Disturbed Vietnam Vet Abuses Italian Boys."

MONDAY. *July 11. I left Rome by train for West Germany, and arrived in Munich for a stay of more than a week.*

JULY 13. *The most beautiful women in Europe collected here in one place on Leopold Strasse.*

JULY 17. *Bought a VW! Landlord suspects I got a bad deal. I bought it at a used car lot after writing down a few German words to help me bargain for cars.*

I named the old Volkswagen Old Gray, and I paid US$350 worth in Deutschmarks. It turned out to be a terrific deal—even a blind squirrel sometimes finds a nut—a great little car that

I drove all over Europe for a year and disposed of at an outdoor café in Andorra for US$550 in cash before returning to the States. I enjoyed Munich, which is probably why I hung around so long. Mostly I remember drinking beer at the Hofbräuhaus and enjoying the sausage and fish street food.

> **JULY 19.** *Drove to Heidelberg. Chock full of US tourists and soldiers. Leaving tomorrow! Gave a lift to a Heidelberg theology student who had answers to everything in the world. Also practiced his orchestral trumpet in the car. He was absolutely no help on Heidelberg.*

> **JULY 20.** *On to Frankfurt, then Hanover, then to the western zone of Berlin after a nervous delay at the Russian checkpoints—probably because I'm an American driving an owned VW. My recollection of West Berlin is a very modern, rebuilt city full of well-dressed citizens and beautiful German women, my meeting or talking to hardly anyone. Most of the people outside were twenty to forty years old. The city seeming insular and closed to outsiders (especially Yanks who were probably Army privates on leave disguised as touring civilians!), with few foreigners around. Wonder how many had dead Nazi dads.*

> *One day I crossed into East Berlin in Old Gray on a day permit and drove through the suburbs where the buildings were still pockmarked with shell holes and rubble remained stacked in alleys. The Russians and East Germans weren't yet rebuilding, and everything seemed deserted. If you wanted to know what the Cold War was about, all you had to do was spend a day in partitioned Berlin.*

> **JULY 27.** *I drove out of West Berlin headed for Hamburg, then on to Copenhagen. Bought Adler portable typewriter for US$42.50 this morning.* [The little Adler made it home and was eventually used by daughter Jeannie in school.]

Went to bed at 7:30 a.m. after staying out all night after dance, driving up the coast looking for a swimming beach, stopping at one, but too cold to swim. Copenhagen railway station—a lot of burned-out Marilyn Monroe-looking women. [Although I stayed there several days, there are no more notes from Copenhagen. I have vivid memories of the enchanting Tivoli Gardens amusement park at night.]

AUGUST 1. *Drove all day up the coast from Copenhagen to Oslo, Norway. Very quiet and enjoyable ride with Linda and Winnie. Not much in the way of conversational company, but in view of the fact that every time they open their mouths they say something stupid or vapid, maybe all's best. Ate a picnic lunch of groceries.* [I retain no memory of these two.]

Oslo. Went to the Viking Ship Museum, Kon-Tiki Museum, Vigeland Sculpture Park, the Fram Museum about Norwegian polar explorations, and the [Edvard] *Munch Museum. A university engineer tells me that Swedes are not liked in his country—they were neutral in the war.*

AUGUST 3. *Saw an okay comedy starring Virna Lisi* [beautiful Italian blond] *and Tony Curtis. Nasty weather, rained most of the afternoon, turned cooler. Went to campus bar later. Short Norse insulted big Norse, big Norse poured beer over his head, and little one punched him in the nose. Linda and Winnie there. Two nice-looking, worldly, German girls, too—one looks and dresses like Karen and is built like her! Declined party invite from Englander because Linda and Winnie would ruin it and I was a stringer to get them tickets to a party night.*

Blustery and cool. Drove down the fjord and took the car ferry across to Moss, Horten, Sandefjord. Linda and

Winnie left from North Norway today—finally gave up on me. They got to be boring and silly to me, especially Linda. Wish I had met Winnie first, although I don't believe she would have been any more attractive alone really, and seemed a bit dense, starved for vivacity, personality. One of those easy-to-laugh, quick-to-depress types. [Wow, I was picky back then.]

AUGUST 6. *Stockholm. Went out with Mayo, a Finnish girl, and another couple to Tivoli. Delightful time. Yet another time remorseful because I'll never see her again. Terrible traffic patterns here, Swedes very discourteous drivers. But they dress fantastically.* [My main memory of Sweden is going into the bar at 10 p.m. or so in the evening as the sun set, drinking with friends until 3 a.m., and upon leaving the bar seeing that it was daytime again. The night had come and gone without me! My other memory is disappointment at the blandness of Swedish meals other than pickled items in smorgasbord spreads, which didn't appeal to my Louisiana seafood taste buds. I'm not big on pickled stuff, and would suffer in Korea.]

AUGUST 7. *Acquaintance with American girl who personifies all I detest, obviously a Manhattanite, perfunctorily rude, abrupt, inattentive, self-absorbed. We snubbed each other! Pissing in the wind again!*

AUGUST 8. *Headed back to Copenhagen. Gave a ride to a post-graduate astronomy type, Bruce. You could pick him out of a police station lineup as the astronomy guy. No enlightenment there, he's in another galaxy. Joan—Mt. Holyoake, her only claim to fame.*

Copenhagen. Received word of John's death. Funeral was weeks ago. [A black day lost in memory.] *Called home. Went to movie this afternoon in desperation to get the sadness out of my head. Picked up letters from Dad, Karen, sister Mary, Annegret.*

REALITY AND HOME INTRUDE

JOHN WAS MY YOUNGER brother who had contracted bone cancer in his knee when he was about fifteen years old and had his leg amputated at eighteen after I left for Vietnam; he died at twenty. The next diary pages while in Copenhagen each have a question mark. I drove on to Amsterdam on the following Saturday.

> **AUGUST 13, 1967.** *Amsterdam. I went bar hopping with Cease and Mary, two Pennsylvania Jewish schoolteachers who were for a change, delightful, expressive conversationists, each twenty-four years old who looked about thirty to me. We were sorry to part at 3:30 a.m.*
>
> *Amsterdam to Ostende to Dover to London. Rained all the way to London. Might as well be driving alone. Brought two English hitchhikers from the Dover ferry into London. Arrived London at 2 a.m. and slept in the car.*
>
> **AUGUST 16.** *Put car in storage, called some English friends, got money and tickets, sent telegrams, and did laundry.*
>
> **FRIDAY.** *August 18. London-New York flight today to JFK.*

Thus ends my final entry in my blue diary. In New York I probably visited sister Mary before flying home to see my family in Louisiana. Although this may be the time I drove the old green Plymouth sedan that Uncle John had given Mary down the East

Coast. I remember getting stopped for supposedly running a red light in Eutaw, Alabama at 5:30 a.m., and giving the patrolman a bribe of my last $25 or so before proceeding to Louisiana.

AUGUST 22–SEPTEMBER 8. *I stayed in Saint Francisville two weeks, longer than I had originally intended; however, I hadn't been home for more than two years. Notes: Swim and hamburgers at Aunt Belle's (Ambrosia plantation). Picked up Dad at Natchez. Went fishing with Alvin Pierpont. Dinner at Farrar's plantation. Saw movie Hawaii. Drinks at Camilla Truax's with nephew Bradley, whom I met up with later in Europe.*

After years away from America, I was appalled by the commercialism, the dominance of materialism in daily life, the relentless advertising on radio and television. Ads seemed to me the scourge of modern existence and killed any lingering vestige of homesickness. Tons of Americans out there still had to strive to live normal lives. Some TV programming is worthwhile as well as seductive. Stripped of all the ads and crap, it could hypothetically be a godsend. How many frustrated souls were tossing everything and leaving NYC and Oakland to homestead and try subsistence living in Alaska?

SEPTEMBER 8–22. I flew back to New York and spent two weeks there before heading back to Europe. Most of my notes in that period were about stock investing, or details of my holdings. I was focused on investing and probably talking to Butch, my brother-in-law, about getting a Wall Street job whenever I came back from Europe. Sister Mary was judging me having morphed into a weirder little soul and leaning on him to encourage me to come home from my expatriate wanderings (and the war?).

Once, I remember landing at the Evansville airport where Dad met me. I hadn't seen him since I left for Vietnam, and I remember hugging him with a lot of emotion, which surprised both of us. I was taken aback by his nonchalance and simple outstretched hand. I must have driven down to Louisiana with him in his car, zero war questions.

At that time, he was spending part of the year in Evansville tending to his well drilling and gas leases. He had bought a house in Evansville, and I suspect his waitress girlfriend from the downtown coffee shop was looking after him. One could say he was leading a double life, something of an escape from my mom's drinking and an obsequious son-in-law hanging around the house in Saint Francisville.

It must have tough on Dad, as well as everyone else, to stand by as John gradually approached his final days. In a sad twist of fate, my father having endured the suffering of my brother's bone cancer and leg amputation, he had one of his legs amputated after contracting gangrene during his long battle with diabetes. Having been a boxer and track athlete at LSU, the loss hit him hard. I stayed a few days, remembering to check out the trunk of belongings shipped home from Saigon the year back in 1966.

PARIS DAYS

SEPTEMBER 28, 1967. *Arrived Paris at 11:15 p.m. after leaving London at 2:30 p.m.*

Getting back to London, I went back to the storage garage where I had left my Volkswagen, finding it covered in dust and broken into but with only my antelope jacket from Turkey missing. Went to see Mitch, staying at his family's palatial empty London flat, shut down for the summer with dust covers on the furniture, etc. Went to see his childhood friend Toby, the dashing barrister who chummed around with us while vacationing down under. Toby was already wheelchair bound with MS. He died within a year or so. Very sad, probably not forty. Mitch, a bachelor, eventually married Toby's widow.

On to Paris. Coming across Asia, I had arrived at the idea that I would use the rest of my savings to live a year in Paris. This idea came from my fascination with literary France and expatriates like Hemingway and Fitzgerald in Paris—along with studying French at Vanderbilt. Remember, I had kicked off on leave in Paris where I got my orders to Vietnam via Western Union telegram. And probably some foot dragging, not looking forward to a return home to an uncertain existence—domestic work drudgery, a more mundane existence ahead.

Arriving on the Left Bank, I spent a few days wandering around. My "home" was Old Gray, my VW. I didn't know anyone in Paris and wandered the streets. Just to be back in Paris and alive was enough; I was happy. Maybe I slept in my car, hostels, or cheap hotels, I don't remember. I found a room for rent with a French family in Boulevard Malesherbes, eighteenth arrondissement. I enrolled in the basic-level French class at the Alliance Française across the river. The classroom's seat and desks were the same wood-and-black wrought iron design we had in high school. Lots of the students were female au pairs or new immigrants trying to adjust to French city life. No English speakers in my class—seemingly all working class, younger than me, and no college graduates.

Still, it gave me a daily routine and the beginning of integration into French ways. With my modest savings, I went down to the Paris office of Merrill Lynch and opened an account with Ron, a friendly Yank expatriate. He was well-connected locally, his spouse the only child of a prominent French-Russian émigré novelist. One night we visited the father-in-law's lavish apartment in the sixteenth arrondissement and had a round of eau-de-vie liqueur, a pear grown in the bottle while on the tree.

Paris was cheap for Americans. I went to the American Express office on Rue Scribe once a month and exchanged the US$500 maximum into francs. That was enough cash to live on for the next month, eating most meals out on the town, and buying a bottle of Rémy Martin cognac every week to heat myself through the night. My room was always chilly as was the rest of the apartment—French families were frugal. Leftovers from lunch usually made supper. In the 1960s, young women in beautiful sweaters probably had only that one or two and treasured them. Memories of the impact of war shortages and occupation were fresh. The traffic was still full of deux-chevaux (two-horse) sedans, rickety little two-cylinder post-war rattletraps that managed to stay in production over forty years.

Paris was exotic and fascinating. I wandered everywhere whenever I wasn't in class. The Left Bank was the place to be for writers, artists, and expats after WWI—that's where I spent much of my time. Although Montmartre on the Right Bank was the bohemian capital of Paris in the late 1800s, the Montparnasse area was a more fun spot. The Latin Quarter held Hemingway's and Picasso's apartments, Shakespeare & Co., Boulevard Saint-Michel, and Saint-Germain-de-Prés with its bars and cafés.

Ron got me an invitation to a party for young European and American expatriates given by a countess on the Left Bank in Rue d'Assas. That's where I met my wife Lisa and her sister Jeannie. I was busy talking pigeon French with a pretty Italian girl who was also learning French and spoke no English. The two Yankee women identified me as the only American in the crowd of mostly women and moved in on my new friend.

We three soon headed out on the town for dinner and became fast friends. They shared a nice apartment off the Champs-Élysées at Blvd. Courcelles in the seventeenth arrondissement while I was renting in more bourgeois eighteenth arrondissement. I think we started dining out nights as a threesome because we were all sick of trying to make temporary friends with expatriates in Paris; it seemed nearly impossible to crack the language and social barriers with young Parisians. We didn't make much of an effort though.

A few weeks earlier, I had been invited to a weekend in the country by the son of my landlord, a group of a dozen French guys and gals slightly younger than me who spoke hardly any English. An empty, cold farmhouse in the Loire Valley. Makeshift meals cooked over the living room fireplace and shared on an ancient wooden table pulled up close to the fire for warmth. Thank God for the abundance of cheap Beaujolais wine! The landlord's son politely served as interpreter the whole weekend.

It was fun. I learned some French but never really fit in. Everyone seemed to know everyone, and no one knew me. They probably went to the same neighborhood school as kids. I suspect my being a combat veteran in the Indochina War and being a little older somehow created a social barrier. Who knows, perhaps a father or uncle had died in the colonial war when France was humiliated at Dien Bien Phu and ousted from Southeast Asia by Ho Chi Minh. Anyway, it was cold and drizzling going there, raining there, and drizzling all the way back to Paris.

In Paris, there was a corner liquor and wine shop across from Lisa's apartment, and we soon became regulars. I set my budget limit at fifteen francs per bottle, about US$3.50. I was learning red wine, and could get either a bottle of Château Lynch-Bages for fifteen francs or two bottles of fresh Beaujolais Villages. Fast forward a half century: I dearly wish I had set my spending limit at twenty or twenty-five francs—oh, the fine Bordeaux Premier Cru vintages I could have tried!

That fall in Paris flew by. School, dining, exploring the city on foot, shopping in outdoor markets, cooking exotic (to us) foods like langoustines, mussels, escargots, or choucroute. One night Jeannie, Lisa, and I ate dinner in a Spanish dive with a flamenco performer. Well into our second or third sangria pitchers, the music and Spanish songs got rowdy, egged on by the audience led by Jeannie and Lisa, who had earlier that year shared an apartment in Madrid. The next song, Jeannie in her cups heads for the stage and dances with the performer. We left there zonked, and I kidded them for days on end for not having brought any money to help with the huge tab.

Another night, we were invited to a dinner party at a gorgeous flat overlooking Notre Dame and the Seine by Lucy, an American liquor-company heiress. Leaving the dinner well wined about 2 a.m., we tumbled into Old Gray and headed west toward home. Passing through Les Halles (farmers market that closed in 1971), I came to a stop in a line of cars at a red light.

We were well numbed so we sat and waited and waited in line to drive on.

After some minutes, Lisa pointed out that we seemed to be stopped at the end of a line of farmers' vehicles in the median of the boulevard. They parked there when setting up their produce stalls. So we, too, parked, got out, and shopped the stalls. Fascinated by the exotic produce, I bought a whole crate of huge persimmons for a real bargain, I think ten francs or US$2! Hauled them home at three in the morning, stowed them in their pantry. Several days later, we rediscovered the persimmons much softer, riper, leaking juice from the odor in the girls' kitchen.

Late that fall, Jeannie returned to the States and Lisa also went home for the holidays. Not me. I had gotten in touch with Brad from Saint Francisville, who was then an executive for Holiday Inns and based in London. Brad was well traveled and fluent in French, and had wormed his way into a holiday chalet lease in Zermatt, Switzerland, by young British friends. At that time, Brits were restricted to US$500 limits on currency carried out of the country, and chalet leases were an end run around severe ski budget limits. I took the overnight train from Gare de l'Est to the Zermatt train station, and a snow sled taxi through the village to the chalet.

> **NOVEMBER 8, 1967.** *Write to Lauri, IRS, Mom.* [My diary pages are full of notes on prices of common stocks I followed or owned. I had started using stop-loss orders. Learning by mistakes, or as Will Rogers purportedly put it, "Good judgment comes from experience, and a lot of that comes from bad judgment." The companies were undistinguished second rank names all of which long ago sank into the backwaters of industrial America. I suppose

their names found their way into my pocket from Merrill Lynch's house "buy" list.]

NOVEMBER 18. *Pound devalued US$2.80 to US$2.40.*

Staying in the roomy chalet was great fun, a dozen or so mostly English girls and us two Yank males, with three of them rotating the detail to cook, keep house, and serve the wine, all included in the entry price. The first day, I went out and bought lift tickets, ski clothes, and rental gear. I had never skied, and in my ignorance and athletic enthusiasm took the lift in my new green down jacket all the way to the 10,000-foot top under the Matterhorn.

Rigging up and plunging downward, I spent the next hours skiing a few feet and falling. Then skiing again and falling—all the way down to the village. By the second day, I knew how to ski—not pretty but no longer falling. I think it never occurred to me that a ski lesson would have been worth the money. My strategy was either challenge or parsimony or stupidity.

Zermatt was great fun, a sparkling snowy wonderland, so I stayed through Christmas and New Year's. No standing watch, no duties, no shooting—just fun. Lisa and Jeannie showed up at some point, and it was wonderful to have my Paris friends there to share the holidays in an enchanting snowbound Swiss town. I couldn't recall ever having three solid weeks of irresponsible fun in a fantastical place, yet I was eager to return to Paris.

While in England the prior fall, I had again picked up a blank onionskin English pocket diary for the 1968 year, and now I made my first entry on New Year's Day, *Zermatt*, then three days later, *Leaving Zermatt for Paris*. At the page bottom, I wrote, *"All women are possessive but some are the devouring kind.* [Source unknown]*"* I was reading history on the train, and made notes that Muhammad died in 632, Muslim armies arrived at Damascus in 635, Toledo in 713, and other such time-lines. Then I jumped into the days of Ostrogoths, Visigoths, Saxons, Teutons, and Vandals, and mapped out where the Gauls, Germanic tribes, Huns, and Romans occupied territory. Saw the movie *Bonnie and Clyde* on the twenty-fifth and wondered if Clyde was a distant cousin.

> **JANUARY 29, 1968.** *Had lunch with three American expatriate women, all rich single heiresses of prominent American industrialists involved in bourbon, gin, and precious metals. One was my contact, a friend of my Kentucky lawyer/writer cousin.* [I joined them out of curiosity and nothing better to do as Lisa had flown back to Tampa. Later we went to a French movie. My contact was lively with a southern woman personality, but the other two were boring, spoiled Midwesterners.] *Kentucky friend had some Russian pamphlets. One busies himself in small moments to compensate for the waste of a life of talent. Lisa returned on Valentine's Day, February 14th.* [Someone wrote a list of several budget restaurant names in my notebook.] *Le Drugstore* [a mobbed retailer on the Champs-Élysées]—*the public Lonely Hearts Club of Paris, always open and brightly lit, and all it costs you is doubled prices and a few friendly shoves in the aisles.*

My note-taking trailed off, but I noted that I was keeping up my French school studies. Although I had now been there for months, I never quit wandering through town because there was so much new and changing and fascinating to see. Of course, my favorite area was always the Left Bank arrondisse-ments. I lunched with Ron from Merrill Lynch, my sort of real-world contact with American business and life. And there were notes about going out on the town with people whose names no longer bring forth events or details.

In late May, Lisa and I got tickets to the French Open and started going to tennis matches at Roland-Garros Stadium at Auteuil on the outskirts of Paris. It was fun, though the weather was cool and often drizzly. The Open, like Longchamp races and Right Bank shopping, were social affairs hosted by Parisians but dominated by international money—meaning rich, jet set, older idle money mixed with the visiting working rich. A lot of these people spent time on the Riviera. Always someone out there bigger, richer, smarter, faster, prettier.

As foreigners, we had social and language barriers just like others staying there. If we clothed ourselves entirely in Galeries Lafayette or Printemps stuff and promenaded down the Champs, we still came across as étrangères. The carefully dressed bodies on the street we took as native Parisians or European jet-setters. They seemed snootier than any New Yorker. The aristocrats of the world were just a bumper ahead of yours and steered like asses and road hogs, worse than any deliveryman or taxi driver.

Meanwhile, all was not well in the real world with the students, peasantry, and proletariat. There were massive demonstrations, general strikes, dumped garbage, and near civil war in Paris and parts of the country. President Charles de Gaulle briefly left the country as police action led to street battles in the Latin Quarter. Lisa and I walked up the Boulevard Saint-Michel to watch the spectacle of students battling armored gendarmes, and ran back toward the Seine River as tear gas drifted toward us.

*Student leaflet hand out at
the Sorbonne — 1960s*

All for Victory
Armed with extraordinary courage and unparalleled tenacity, the people of Vietnam inflict stinging
defeats on American forces every day......Johnson undertook the assassination of the entire civilian
Population.....we will prevent this genocide....the time for direct confrontation with U.S.
representatives in Paris will come...today our duty is to make Feb 21 the first step...side by side,
Students, workers will occupy the Latin Quarter and demonstrate from 6:30 pm...So that the FNL
becomes the power in Saigon...for Americans to be thrown out of Vietnam.

TOUT POUR LA VICTOIRE

Armé d'un courage extraordinaire et d'une tenacité sans pareille, le peuple
vietnamien inflige chaque jour de cuisantes défaites aux forces américaines.
Les combattants font la démonstration de leur maîtrise militaire et de l'im-
mense soutien populaire dont ils bénéficient. La preuve est faite qu'il n'exis-
te plus un seul endroit sur le sol du Vietnam où l'agresseur puisse s'instal-
ler impunément. Battu sur le terrain; Johnson entreprend la destruction sys-
tématique de villes entières, l'assassinat délibéré de toute la population ci-
vile.

NOUS DEVONS EMPECHER CE GENOCIDE
LE COURAGE DU PEUPLE VIETNAMIEN EXIGE DE NOUS
UNE MOBILISATION DE TOUS LES INSTANTS

L'heure n'est plus aux manifestations traditionelles se limitant à la simple
expression verbale de notre indignation. Aujourd'hui le gouvernement gaul-
liste met en place devant l'Ambassade Américaine un dispositif policier con-
sidérable pour empêcher toute manifestation. Nous ne pouvons être comp-
lice de cette reculade devant l'impérialisme américain.
Le C.V.N. n'abandonne pas la perspective d'une manifestation devant l'Ambassade. L'heure de l'affrontement direct avec les représentants U.S. à
Paris viendra. Elle sera d'autant plus proche qu'un très haut niveau de com-
bativité, d'organisation et de rassemblement des forces sera atteint.
Aujourd'hui notre devoir est de faire du 21 février une première étape dans
le développement d'actions sans cesses grandissantes qui nous permettront
de passer outre à toutes les interdictions.
Le 21 février, depuis des années journée de lutte contre le colonialisme et
l'impérialisme, sera pour nous l'occasion de manifester d'une manière ré-
solue notre soutien au Vietnam Héroïque, notre volonté de voir se rapprocher
la défaite des agresseurs U.S., notre détermination de ne plus borner notre
action à de simples défilés protestataires.
Côte à côte, étudiants, lycéens, travailleurs occuperont le Quartier Latin
et manifesteront dès 18 heures 30 sur le Bd St Michel. Le caractère unitai-
re de l'appel de l'U.N.E.F. donnera à la manifestation une ampleur qui dé-
passe le cadre du mouvement étudiant.
Nous ferons du 21 février une journée pas comme les autres. La journée la
plus claire, la plus décisive, de notre soutien au Vietnam Héroïque. La journée qui permettra à de plus larges masses de prendre conscience de la résolu
tion des anti-impérialistes. La Journée qui obligera tous les partisans d'un
"compromis honorable" à jeter bas le masque, à reconnaître qu'ils refusent
d'admettre la signification politique et militaire des récents combats.

POUR LA VICTOIRE DU VIETNAM
POUR QUE LE F.N.L. SOIT AU POUVOIR A SAIGON
POUR QUE LES AMERICAINS SOIENT JETES HORS DU VIETNAM

TOUS AU QUARTIER LATIN LE 21 FEVRIER, 18 HEURES 30, BD ST MICHEL

Comité Vietnam National, 22 rue E. Marcel. Paris 1er.

Paris student protests-Vietnam

I was curious, amused, and perhaps a little alarmed. This upris-
ing could be history in the making, so we stood and watched

and wandered the Left Bank. As a bit older war vet, I had little respect for the milling disorganized students searching for excitement and trouble under the direction of leftist leaders. I'm not a participant sort for "causes." In most cases, about half are on one side, and about half are on the other. That means about half are right or everyone's wrong.

> **MAY 30, 1968.** *Pro-Gaullist parade down the Champs-Élysées. Gas is being sold again, so the strike is over. Finished my French studies—third degree—at the Alliance Française.*

> **JUNE 4.** *Robert F. Kennedy assassinated.*

> **JUNE 12.** *Lisa and I celebrated my twenty-eighth birthday at the luxury restaurant Laserre, under the stars as they slid open the roof midway through the courses.* [I recall being fascinated, then afterward mortified, at spending 550 francs for our meal and wine—about US$100 equivalent those days. I was enjoying life on my $500 a month budget in Paris and that splurge blew a hole in my budgeting. Perhaps I had already realized my carefree expatriate days were winding down.]

> **JUNE 15.** *My Volkswagen got hit in the fender.* We had been out on the town that weekend on foot/taxi/Metro and I had left Old Gray parked on Boulevard Courcelles. When I went Sunday morning to retrieve it, the front fender was scraping the pavement, nearly torn from the car. I looked around Paris for a repair shop and discovered that it would cost nearly the VW's market value to fix it. Shortly thereafter, we hauled out to tour France and Spain, steering a front-right bumper and fender stitched together with baling wire and pliers purchased at the hardware store.

JUNE–AUGUST 1968. My notebook days are filled with itemized traveling costs as we motored down to northern Spain. We first drove west to La Rochelle on the coast. I remember trying my first oursin (sea urchin) dish at a seaside café and concluding I had found one seafood that requires an acquired taste in order to ever possibly enjoy it. Down through Amboise, then Biarritz, and over to Burgos, Spain.

Near Burgos, we found a garage to make the repairs, which turned out fine. They cost about a fifth of repair estimates in Paris. We went to a local bull fight, far more authentic than the knockoff skirmishes I had seen in Mexico and South America. I had long since read Hemingway's book *Death in the Afternoon*, about his experiences with the bullfighters' world while living and writing in Spain. Jeannie and Lisa, inspired by living in Madrid, had rhapsodized about matador life.

We drove down through Toledo, then on to Madrid, and west to Coimbra and Lisbon. A small restaurant on the coast of Portugal specialized in grilled chickens spitted whole on a contraption turning the spits stacked horizontally one above the others. The waitress would go over and move the spits to different positions every few minutes, basting the top one with lemon, olive oil, and garlic, which then dripped down onto the lower chicken. The best, crispiest chicken I had ever eaten. I suspected those skinny birds had subsisted on free-range worms, bugs, and seeds.

On we drove, down the coast to Lagos, Portugal, and picnicked on big sardines grilled on a little beach fire among the spectacular boulders abutting the ocean. Beautiful beaches, a seafaring port. A few days later, we headed back into Spain across Granada, Valencia, Sitges, and to Barcelona. Bumping along on two-lane roads with no air conditioning, the little VW turned into a frying pan in the relentless afternoon heat of near-desert terrain. But Old Gray never let us down. From her Spanish studies, Lisa had a good understanding of medieval architecture dating to the Moorish occupation and became my tour guide to the Alhambra and its beautiful palaces and gardens.

We came back across to Andorra. I asked around town for a way to sell Old Gray; by noon, I'd met a buyer at a table outside a café. Over two beers, he paid me US$550 cash, no questions asked, no papers needed. That was $200 more than I had paid in Munich the previous year. Oh, the stories that VW could tell!

With the trip winding down, we took an uneventful train ride back to Paris through the southwest French countryside. Wistfully looking out at the passing scenery, I was thinking of things coming to an end. A few weeks later, I left for New York and home. Adventures ended. It was nearing time to enter the ordinary world, look for a job, and try whatever else.

COLLEGE AND NAVY TRAINING

IN THE POST–WWII 1950S, the conventional route to adulthood after high school was college or a job. If still single, teens often went straight into a local full-time job, doubled up with local schooling, or entered the military. College students who could afford it attended for four years until graduation, or left for the work force after two years at a local college. The Korean War interrupted the return to normality for many GI vets, but the early Cold War years saw the expansion of officer training as an option for undergraduates at a few dozen top universities.

My older sister Mary attended Vanderbilt in Nashville, and I followed her there. University was my second choice. My first choice was the chance to be a deck hand on the Miami yacht of an oilman in Evansville who kept a cabin cruiser on the Ohio River. I met him while working at the Evansville boat club. He told me stories about winter cruising in Florida and the Gulf of Mexico. He wanted to take his yacht to the Mediterranean, and I told him I would work for him for nothing just to get out to sea. As it turned out, my parents did not think much of the idea, preferring the conventional route of my following in Mary's footsteps.

Mary thought my following her was a smart choice. She carried some of her demons from home to Music City, but got rid of a lot of others there. Mary was the only neat and tidy member of our household. I was persona non grata in her bedroom, her inviolable sanctuary. She had an artistic bent like me, drawing and painting in her room. In school, she tested at the genius level on the IQ exam. She was independent, opinionated, and hardheaded in my younger brother view, and was a bit absent-minded thing. Mary would forget to get off the bus or train, lose keys, and leave her books behind.

She graduated in four years and went on to marry and have a happy family and social life in New Jersey. She always had a few close, highly loyal friends starting in grade school. She kept a somewhat-secret cigarette habit. She was diagnosed with lung cancer at sixty-one; she tried an experimental drug for several months. One Monday morning—after being pronounced cured the previous week—she lay down on the living room couch and died. We were always different but still close; we went through a lot together.

Vanderbilt heaped the social nonsense of freshman rush on top of new study routines and a frenetic, noisy dorm scene. I made the rounds of fraternity houses and finally pledged Kappa Sigma, the second or third largest of about a dozen fraternities on campus. I was playing the college game, but didn't fit into the mold of fraternity social life with designated buddies. Down the street, there was the enticement of an unfamiliar city—chock full of music hangouts, bars, and cafés. Tennessee had lakes and rivers with fish species I had never caught. And livelier dates across the street at Peabody College.

In college, I quickly learned that there were smarter people than me, especially regarding study discipline and routine. As a school boy, I was one of the smartest in class. My only competition was Bruce, who rarely made mistakes on tests. He was plugged in to assembling ham radios by the age of ten.

At Vandy, it was a shock to start feeling a bit dumber. I came to view the classes and social activities as simply a way station to the future. Back then, some people knew they were average, and some knew for sure they were slow learners. Today, most people, with Google at hand, think they're pretty sharp. Everybody has the answer from the start, or can find it and testify to your mistake. Never have so many known so little about so much.

Most of my college-bound high school friends stayed in Indiana. At Vandy, I knew only my older sister and a guy in her high school class who had come down from Evansville. Two of us studying out of state at a relatively pricey private university soon became a financial burden to my parents, though this was their own making. I got word that Dad thought Evansville College—a mile down the road where my best friends studied—would suffice for me. Despite my lack of true love for Vandy, hometown schooling was an unsavory thought.

I joined Naval Reserve Officers Training Course (NROTC) and enjoyed the mandatory freshman naval history and weapons courses—not so much the dress parades and petty discipline. Shooting matches with the NROTC rifle team were fun. I tried out for the swim team after the swim coach saw me swimming and asked me to join practices. The pool chlorine, however, irritated my eyes and I quit after a few weeks. Most the team had been competitive swimmers in high school, not me. Literature and history were enjoyable. Art history was a tedious farce ladled out in a cramped little ex-chapel by a portly gay man who adored Gothic, Ionic, and Corinthian styles.

In my sophomore year, I managed to pick up a Navy scholarship providing me room, board, and books—plus the munificent sum of $50 a month for expenses. In return for switching from reserve to regular status, I became the Navy's indentured servant for two additional years (a total of four). I had to make up the freshman summer active duty cruise I'd just missed. Navy courses, one each semester, continued through my senior year.

Most of my summer of 1959 was taken up by the first of four tours of active duty. As a one-stripe midshipman, I was sent to New England to meet my duty ship, spending it in the North Sea aboard the USS *Abbot* (DD-629) a WWII-era destroyer. Destroyers were nicknamed tin cans as they were thinner armored and flimsier than ships of the line like cruisers and battleships. We had the honor of bedding down in the forecastle compartment just aft the ship's anchor locker. On calm nights off duty, our bunks would gently rise and fall in never-ending twenty-foot cycles reminiscent of a county fair ride.

In heavy weather, we would brace our knees, elbows, head, hands, and/or feet against the bulkhead, the bunk above, or pipes overhead to keep from flying out onto the deck—and pray for sleep. We managed to catch a freak seventy-foot wave summer storm off Nova Scotia and pulled into port for painting and repairs. While refueling off a tanker out at sea, a huge wave between the ships split the fuel line, engulfing the starboard side of the ship in greasy bunker fuel. Two sailors were lost over the side. At sea, bad stuff can happen quickly. I never forgot that early lesson.

The town of Sydney, Nova Scotia—on a wan and rocky seaside island—hosted a dance for us midshipmen, attended by most of the teenage girls in the small town. I recall my dismay at their tea-stained smiles. If there were any pretty girls in town, they didn't show that night. Later, it occurred to me that some of these parents from seafaring islands may have taken snot-nosed foreign midshipmen as the equivalent of box-car rounders passing down the track through town. Nevertheless, we were the first American warship to visit Sydney since WWII. The occasion called for gala hosting, with a parade celebration, speeches, boat races, and a warm reception by the orange-toothed maidens. Six of us midshipmen in our rescue boat outoared their matching boat in a mile race across the harbor. I was a conquering hero for thirty minutes.

Those summers on active duty as a midshipman were adventurous new experiences getting to know Navy life and new places. The sailor uniforms and hats were goofy and embarrassing in the beginning, but the officer uniform looked manly. Going on liberty in such outfits cut both ways, limiting our range of social moves but often attracting silly young girls.

All the new rules and regimentation were an aggravating grind. In the military, you learn right away there are two kinds of people: those who leave their toothpaste open, and those who put things back where they found them. We all got forcibly molded into the second group.

I never had thoughts of becoming a career admiral or Army general. Other boys dreamed of becoming fighter pilots, building and hanging balsawood model planes in their bedrooms. My interests were broader and vaguer. My instincts were to become an accomplished generalist, not a specialist. Despite my woods skills and admiration for boy heroes like former President Teddy Roosevelt and American trapper Kit Carson, I was a surprisingly poor Boy Scout, silently railing against regimentation and groupthink. Military doctrine was usually the product of armchair, not battlefield, generals. In the Navy, a lot of the fun was in exploring new ports and cities, in fleeting friendships, and blind dates with girls wherever we happened to be. Climbing the career chain? No interest.

My next three summers were spent at flight school in Pensacola and Corpus Christi, Texas, aboard the nuclear submarine USS *George Washington* (SSN-2) out of New London, Connecticut, at basic grunt Marine training in Little Creek, Virginia, and aboard the USS *Kitty Hawk* (CVA-63) a nuclear carrier based out of San Diego and cruising out of Honolulu. The Navy was introducing me to its different parts as their way of fitting us where we would best perform.

I enjoyed flight training and performed well, but never wanted to pilot planes. We put our swimsuits on, climbed into a cockpit mockup, and got dropped into a fifteen-foot water tank where we had to throw back the canopy, release seatbelts, and swim to the surface. We then studied and flew T-34C single-engine trainers making touch and go landings with our flight instructor. In Texas, we did the same in two-seat T-28 jets, and did loops and rolls to test how our body and brain handled the stress of g-forces. No picking puking, pooping, or passing out pilots for advanced flight training.

Ship/vessel: USS Abbot (DD-629)
Type: Destroyer
Year trained or served: 1959
Length: 376'
Image source: Naval History and Heritage Command

Ship/vessel: USS George Washington (SSBN-598)
Type: Ballistic missile nuclear submarine
Year trained or served: 1960
Length: 382'
Image source: Naval History and Heritage Command

Ship/vessel: USS Kitty Hawk (CVA-63)
Type: Nuclear aircraft carrier
Year trained or served: 1962
Length: 1,069'
Image source: NARA & DVIDS Public Domain Archive

Ship/vessel: USS Suffolk County (LST-1173)
Type: Tank landing ship
Year trained or served: 1963–4
Length: 446'
Image source: Wikipedia

Ship/vessel: USS Georgetown (AGTR-2)
Type: Technical research ship
Year trained or served: 1964–5
Length: 441'
Image source: Naval History and Heritage Command

Ship/vessel: Yabuta Junk
Type: Diesel coastal raider
Year trained or served: 1965–6
Length: 36'
Image source: National Archives

Aboard the nuclear submarine, we traveled quietly; we trained to live quietly. We often crabbed sideways through compartments, stepping and stooping through hatch doors spaced closer than on surface ships. In one training session, we were shown into a compartment entirely devoted to housing a huge IBM computer devoted to continuously calculating the bearing and range to targets like, say, Moscow. The next day we rose to periscope depth and tested our Polaris missile tubes, which forced the missiles above the surface with compressed air. I was in the Combat Information Center (CIC) and watched a Russian trawler a few hundred yards away witnessing the test, dozens of Russians on deck manning binoculars and cameras.

The Marine business I remember best is hiking with a backpack, doing obstacle courses, and marching—all of which must have excited the competitive grunt drive in aspiring Marine officers. The carrier cruise in the Pacific was a fun trip aboard a floating city. A surface ship, a line officer—that's where I belonged.

Being a Navy midshipman meant wearing officer's uniforms— work khakis, dress whites, or dress blues—except that rank stripes were replaced by thin yellow bars branding us as novices. We were respected in public and snickered at in private— budding junior officers-to-be. That was our public face, but on duty onboard, we wore sailor's dungarees or Marine-enlisted men's greens. We invested quite a few hours swabbing decks with wet mops as deck department seamen stood by watching. On the other hand, college girls loved our uniforms and often knew that we came from elite colleges, including most Ivy League schools. We strolled down the street on liberty and readily met the local girls.

Back at Vandy, I tolerated my junior and senior years, but I was aching to hit the road. I lived off campus and dated a nice working girl from a broken family on the south side of town—she was just out of high school. Her mother was receptive to me, worked hard, and tried to avoid us. I found excuses to avoid studying. When lacking anything better to do, I wandered down to my frat house or went to one of the house parties. My major was economics, but my love was literature and I took as many credits as possible in the English department. My favorite course was Modern British and American Fiction. When exams approached, I crammed for the tests since classes were usually not challenging for me. I spent little time studying. Going down to Printer's Alley, the Grand Ole Opry, or driving to Center Hill Lake to fish were the best of what school life offered me.

One of my fondest memories of my last year was drinking and listening to reel tapes of *The Student Prince* into the early morning hours with a sweet blond and her friends from Peabody

College across the street from my campus. I was heading for the real world. We had nothing in common except loving the music and drinking. She was pretty, nice, pleasant, and liked me. I went for her taste in music and the laid-back ways of her friends. That lasted a few weeks before the fade and we mutually moved on.

I graduated in June 1962 with a bachelor's degree in economics and minors in international affairs and literature. I never attended a Vanderbilt class reunion. My parents weren't there for graduation. I had my little English Ford all packed with my luggage when I went up on the stage to receive my diploma. The negative of having a last name starting with B in the classroom finally paid off, and I was soon driving out Twenty-First Avenue headed for Louisiana. Through the windshield, I could see the stage across the street where the chancellor was calling out the end of the alphabet. No tossing of caps or hugs all around for yours truly.

First, however, I had to report to San Diego for my make-up midshipman summer cruise aboard the USS *Kitty Hawk*, a monstrous nuclear carrier based on the West Coast. We had the pleasure of a couple of weeks to explore the town and get in trouble with sailors partying in Tijuana before setting sail for Hawaii. In Honolulu, I had my first-ever gin and tonic sitting under Robert Louis Stevenson's giant banyan tree in the courtyard of the Moana Hotel looking out at Waikiki Beach on a full moon with young ladies giggling at us from the next table.

The ship shortly took us out to sea on night operations. From a wing of the flying bridge, we watched as F8U pilots practiced takeoffs, approaches, and landings in the pitch black. The F8s had one Pratt Whitney engine whereas the F-4s had two, so carrier landings for the F8s had no room for error with only half the thrust. We lost two planes, which ditched off the side into the blue when they missed the landing deck. The second plane's pilots ejected from the cockpit sideways, straight into the waves, and were killed instantly.

SEE THE WORLD, FIND A WAR

FLASH FORWARD THREE MONTHS. After finishing my make-up Pacific cruise, I slogged back to Nashville one final time in the dead heat of August across an empty campus for a five-minute commissioning as ensign, US Navy, by a shore-duty recruiting officer who drove out to campus from downtown. As the last person in my class to be commissioned, I was last to request a duty assignment preference. I got the "choice" of whatever was left, but by luck it worked out. I was soon shipped right into the Cuban Missile Crisis with operations in the Caribbean thanks to my new duty in the Atlantic amphibious fleet.

My first ship, the USS *Suffolk County* (LST-1173) lay stranded in dry dock on the Saint Johns River in Jacksonville, Florida. She was in the shipyard undergoing refitting. After a few weeks of easy duty, learning my way around the ship and the fun spots on Jax Beach, we set sail north to our home port, Little Creek, Virginia. I was assigned as deck and gunnery officer, but since we were primarily an assault landing ship, my main work involved the leading two dozen seamen in deck division. Swabbies they were called, as they spent countless hours swabbing the main deck and passageways with mops, suds, and water. Twenty-two-year-old Navy ensigns at sea were learning their jobs while trying to lead more seasoned enlisted men who already knew their duties. Similarly, my subordinate petty officers were generally older and more experienced in shipboard ways than I.

The newer deckhands were a motley crew of everything from kid recruits to crusty old underachieving seamen. They went to sea under the care of mother Navy—maybe to escape home problems, the law, the bottle, or just to be at sea for the adventures of military life sailing to new ports. A gal in every port. I needed to sort out these guys I was suddenly bossing. Some were dependable, experienced, trustworthy. Others as raw and new as me, or simply troublesome slackers. Be careful judging people by their looks: some of the most memorable people I met looked like store clerks or municipal pothole fillers. We had an old seaman with a master's degree scrubbing decks. I traded paperbacks with him weekly.

While in port, sailors were always angling to get off the ship. The anti-aircraft tracking radar for our 3"/50 turrets seemed to go on the blink every couple of months. Jones, our gunners mate first-class, would get permission from the XO to hand-carry something in the radar's brain called crystals to the factory to exchange for a new pair. As I recall, replacements cost ten or twenty thousand dollars. Off would go Jones, Korean War veteran, in his crisply-pressed dress blues and mirror-polished shoes, a twinkle in his eye.

A few days later, back from the factory somewhere up in central New Jersey, he would arrive cradling a little box of new crystals. He would install them, and grinning, report the gun turrets ready-to-fire to me, his immediate superior. I would say something like, "Yeah, and nice trip to town, eh?" I smelled a rat, and reckoned there was a girlfriend somewhere near that Jersey town- perhaps even behind the factory sales desk. I can see my crusty old gunner's mate, pipe in mouth, years later saying to the other Chiefs, "I bet that Ensign kid was on to my crystal trips the whole time."

One day as we were undergoing a final inspection critique in the officer's wardroom, our work was interrupted by an urgent communication from the Pentagon. Our entire squadron of ships was ordered to get underway in four hours. That night, as we sailed south to Onslow Beach in the Carolinas, the ship's speakers broadcast President John F. Kennedy's speech telling the world that the USSR was shipping guided missiles to Cuba. Our Navy would blockade the island to prevent the missiles from reaching land. The next morning, we loaded an entire Marine Corps battalion of 1,050 combat-ready Marines onto the ship along with provisions and Amtrak amphibious tanks in the cargo hold. The Marines mostly slept topside, under the stars.

At sea, everyone on board looked forward to the next mail call or exchange of movies if there was to be no port call in coming days. During the Cuban crisis, there were no port calls, no liberty calls. Mail or movies for our ship came from the lead ship in the squadron, a landing ship, deck (LSD) with a helicopter deck that received mail and movie reels by air when within range of land or other squadrons. The movies and mail would also come over to us by highline while refueling at sea. One reel would be exchanged for one of ours already watched. We never knew whether we were getting a good exchange or just some old shit-kicker—an old generic western, already passed around the fleet several times. Gary Cooper, John Wayne, Jimmy Stewart, Randolph Scott, Alan Ladd, Lee Marvin, and Clint Eastwood became hero uncles. Even Tom Mix and Gabby Hays. For the next three months, we sailed through the Caribbean crammed with Marine grunts from stem to stern. Hurry up and wait, watch, gossip, sleep, read, think, eat, poop.

When there was no fresh movie to watch at sea, the idle hours
not on watch, on operations, or sleeping were spent reading,
playing cards, or jawboning. I imagine that sailors in the old
days spent leisure hours at sea playing cards and games, singing, jawboning, and reading the Bible or other books if they
were literate. Perhaps they carved wood or ivory, or knotted
and repaired lines and hammocks, and practiced whatever
other craft they knew, or engaged in close-quarters sports like
arm wrestling—particularly if a grog ration was issued on their
ship. Those were different times, ships, sailors, and uniforms,
but it was the same seas and the same daily routine altered just
by time—minus the grog ration. When the ship's store was open
and we were in international waters, I could buy my beloved
Pall Mall cigarettes for ten cents a pack, one-third the cost back
in port. Most of us smoked and drank coffee like mad.

Sailing under wartime conditions with our onboard population, quadrupled by a topside deck festooned with lounging
leathernecks, crimped our daily routine. As a deck officer, my
workload doubled, but we still had to make time for drills: general quarters, fire, man overboard, antiaircraft, repel boarders,
abandon ship. We finally docked at Naval Station Mayport, just
north of Jacksonville in January. A few weeks later, the crisis
had abated and the battalion of caged souls debarked, headed
for their military home in North Carolina.

We returned to the Caribbean, cruising offshore from Puerto
Rico, Cuba, Jamaica, and the American Virgin Islands. Going
on liberty in San Juan, Charlotte Amalie, and Montego Bay
was a great adventure for me. Guantanamo Bay, Gitmo Naval
Base, not so hot. San Juan had become the tourist and expatriate center of the islands after Havana shut down. Even Bacardi
moved its rum manufacturing plants there along with Havana
girls, gamblers, and gangsters. I loved hanging out at the Caribe
Hilton bar and playing roulette at the casino. Lots of whisper-

ing and intrigue were about even if we weren't in on the latest scoop.

Out on one patrol, the entire squadron lined up in column formation off Vieques Island for gunnery practice. The target was an old WWII destroyer escort, long retired from service and mothballed, anchored about 5,000 meters to sea. As gunnery officer, I took my place on the flying bridge. The USS *Suffolk County* got first shot at the head of the single file column of ships because we had the lightest guns, twin three-inch/50-caliber antiaircraft turrets on the bow and a single turret on the fantail. I told my crusty old Korean War chief gunner's mate to load armor-piercing shells. When we cruised into range, the squadron admiral gave the order to commence the exercise. Our initial shot hit the water 50 meters in front of the ship; the second penetrated the hull at the waterline.

Not waiting for word from my captain, I ordered my chief, "Commence rapid-fire!" He threw the turret into antiaircraft automatic fire, slamming round after round into the hull. In less than a minute, the escort's bow rose high out of the water and it slid stern first beneath the sea—like in the movies. The admiral was yelling, "Cease fire, cease fire" through the ship's speaker, and my captain ran over to me repeating, "Cease fire, cease fire!" My chief had already ceased fire because there was nothing left floating, and he and his crew were already out of the turret cheering. We sank a ship! My gunnery crew was ecstatic!

The entire squadron of thirteen ships following us had nothing to shoot at. The exercise was over in a minute, and shortly thereafter the admiral was chewing out my captain. Next was my turn. He came marching to my side of the bridge, but I could see him grinning. He scolded me that we weren't supposed to hog all the action, then gave me a "well done" and that was it. By the next morning, the port gun turret had a 45-degree sinking ship painted on its side. When we entered port for liberty, the captain ordered it painted over.

Our home port was Little Creek, Virginia, just outside of Norfolk. When we returned to port we navigated through one of the world's the most congested shipping channels. On my last return to port there, I had the mid-watch (midnight to 4 a.m.) on the bridge, taking the ship into port by radar on a black night—steering through shipping traffic by surface search radar. As the officer of the deck, my watch rotation made me senior officer in charge while most of the crew, and captain, slept. I was responsible for assuring that the watch crew of a helmsman at the wheel, a petty officer at the radar, a signalman, and look-outs with binoculars and headphones posted on the bow and stern safely navigated the boat into port.

The flat round glowing screen of the surface search radar showed blips of moving ships in a circle radiating out from us, the source, at the center. The radar beam, a line on the screen from the center to edge, swept clockwise like a minute hand and refreshed the position of each ship's blip every thirty seconds or so. With a grease pencil, we could mark the position of any ship on collision course and the next sweep would recon-firm its approach angle. It was easy to mark the course of a few approaching vessels in a channel, but the confluence of US naval vessel traffic and numerous commercial freighter, tanker, fishing boats, and private boats meant that our radar screen quickly became cluttered with grease marks, tracking potential collision courses with various yet unseen vessels. A collision course was a decreasing range (distance) on a constant relative bearing. I made several course changes and sent several flash-ing light signal warnings by the time I passed into calmer waters and finished my four hours of watch dripping in sweat. Fifty years later, I still get shivers thinking of bringing my ship into port in Virginia in the very alive "dead of the night" while the captain and crew dreamed of mermaids.

I loved Little Creek because down the road a few miles from the base sprawled Virginia Beach. What do bachelor Navy ensigns do on liberty? Go to the beach. Bars, seafood cafés, music joints, bachelor schoolteacher gals sharing beach house rents, weekend house parties with blaring music up and down the beach. I remember one weekend (back from a month at sea), we holed up in our rental house with Tennessee schoolteachers as a near-miss hurricane passed. It became frozen daiquiri weekend. We had plenty of spirits, ice, and a good mixer, making daiquiris out of every fruit and sweet in the kitchen. By the second drinking day, as the rain poured down in torrents, and Joan Baez, the Beach Boys, and Peter, Paul and Mary blared, we got down to emptying the pantry.

The good mixes like grapefruit, orange juice, and limes ran out and we went through the fridge and pantry stock of canned fruit. The grog became canned peach and fruit cocktail frozen daiquiris downed in a haze of pounding rain with Harry Belafonte crooning on a seventy-eight record. We may well have downed the only canned spinach daiquiris in history. It was a new adventure, we were young and independent, doing things with people we hadn't known weeks before, who like us were living just for this day.

In January 1964, we got called away again on short notice to haul another battalion of Marines south. A flag-flying school dispute in Panama had erupted into street fighting between citizens and Canal Zone soldiers, with deaths on both sides. We picked up fully outfitted Marine combat troops at Onslow Beach, North Carolina, and sailed down to Panama, passing through the canal to the town of Colón on the Pacific side.

Restricted to mooring alongside a basketball court on the water's edge of a city park, we sat and waited, loaded down with antsy grunts day after day as nothing happened. Finally, as it looked like war wasn't going to erupt, I got permission to take a rickety old train across the isthmus through shaded green jungle back to Panama City for a shopping trip. I arranged for bingo games, popping corn, and dozens of cases of beer to be trucked back to the ship from the Army post exchange (PX) store. The basketball court was ringed by a high steel fence and we got permission to give the troops a night off the ship swilling beer, eating popcorn, and playing bingo on the asphalt court until midnight. After three weeks aboard ship, it was a massive celebration—nearly one thousand beer-drinking, popcorn-eating and bingo-playing soldiers with blaring rock and roll on speakers. We headed back home through the canal a few days later.

After two years duty in the Atlantic and Caribbean, it was time for rotation to a new duty station. I worked a connection at the Pentagon and found an opening for independent steaming on the USS *Georgetown*, an electronic surveillance ship also based out of Virginia. The *Georgetown* was one of three naval vessels outfitted for Cold War offshore spying by intercepting and translating communications traffic. (A sister ship, the *USS Liberty*, was attacked by Israeli naval forces with thirty-five US sailors killed three years later during the Six-Day War). Working my accumulated leave to miss the ship's departure, I caught an Air Force C-130 cargo plane leaving for South America from a South Carolina base with a load of cargo under canvas tarps. We took off one morning, me alone on the canvas side bench facing a strapped-down stack of mystery cargo in the center, with pilot and co-pilot up in the heated cockpit.

I next saw them eighteen hours later when we landed for refueling in British Guinea on the north coast of South America.

I finished my two paperbacks and was battling near-freezing temperatures in the plane's cargo hold. After refueling in the rain at a steamy jungle base, we took off on our final leg for Rio de Janeiro, the *Georgetown's* first destination port in South America, after crossing the equator.

My clever scheduling had worked for once; twenty-four months learning the bureaucratic ropes as a junior naval officer and I finally hit pay dirt. We landed in Rio fully ten days before my ship pulled in to port. I was getting my full pay, plus a per diem allowance and temporary duty pay with no duties and nothing to do. With Brazil in the throes of hyperinflation, a US dollar bought me 650 cruzeiros of their currency. I loaded up and headed for Ipanema Beach and the sights. This happened to be the spring of the coup d'état from democracy into dictatorship for Brazil, and also the release of "The Girl from Ipanema" song, soon to become a hit worldwide. Of both, I was oblivious as I passed through town.

Rio was beautiful, sunny, and fun with no sailors and few tourists in 1964. I sunned and swam at the beach, marveled at the bikinis, and drank fruit drinks and cachaça, the Brazilian sugarcane rum. I walked the town, took a tour bus, and tried feijoada, the national dish of black bean stew loaded with meat and sausages. Somehow I ran into Ron, a fellow Yank who had cashed in the return trip plane ticket to Rio rather than using it to go home; his parents had given it to him as graduation present from University of Miami. He stayed in Rio and never came home.

Ron got together with an ex-Nazi pilot and they hired out to aerial map the interior of Brazil for the government, opting to take half their pay in hectares of jungle. Their pay quickly translated into huge blocks of virgin interior. By the time I met Ron, he was living in a penthouse on Copacabana Beach and had just hosted Playboy magazine's "The Girls of Brazil" photo shoot at his digs. I showed up in time to stay a few days in his penthouse as things were winding down. Magnificent beach views, leftover party supplies stacked in corners everywhere. A couple of days later, the glass slipper broke, the party was over, and I headed for the port by taxi to meet my ship. In little more than a week, the cruzeiro had crashed to 1500 in the Brazilian inflationary spiral, doubling my money and paying for my bacchanal week.

My new ship was great duty. We cruised alone down both coasts of South America and through the Caribbean islands on Cold War electronic surveillance missions. The *Georgetown* bristled with antennae, radars, and other gadgets. Below decks, the entire center of the hull was cordoned off by two bank vault doors with duty watchmen restricting entrance to those with proper security clearances. As gunnery officer and assistant navigator, I never stepped into that part of my ship. It was none of my business. Most of the crew had communication technician (CT) ratings, with foreign language and radio gear specialties and top secret security clearances for their eavesdropping work.

My first week on board, we went to sea for some housekeeping duties. On the ship we carried a cargo of dozens, maybe hundreds, of cases of mouthwash with defective labels. As I recall, it was donated by a bottler to a UN agency, which convinced the Navy to give it away in South America. Our captain, a sensible guy, had the bottles brought topside, the contents poured into the sea, and the containers crushed and trashed. Then we headed south to Uruguay, and up the Rio de la Plata to Buenos Aires. My first memories of Buenos Aires are hazy, because I returned to Argentina several times. I recall steering past the vast schools of small penguins leaping in unison on the horizon during my watch as officer of the deck.

The wardroom stewards, or officers' cooks, were nearly always Philippine sailors or Black American in those days. They always delighted in preparing a meal more interesting than standard Navy provisions requisitioned from the supply officer. They used their creativity and improvised, sharing any specialties they came up with. The wardroom, the officers' dining room, was our special private refuge when off watch. We met there, ate meals, had meetings, played bridge or acey-deucy, watched movies when available, or hung out and drank coffee.

IMPERIVM NEPTVNI REGIS

To all Sailors wherever ye may be:

And to all MERMAIDS, WHALES, SEA SERPENTS, PORPOISES, SHARKS,
DOLPHINS, EELS, SKATES, SUCKERS, CRABS, LOBSTERS AND ALL OTHER
LIVING THINGS OF THE SEA GREETINGS: / Know ye: That on this
27th day of January 1965, in Latitude 00000 and Longitude
81° 00'W there appeared within Our Royal Domain the USS
GEORGETOWN (AGTR-2) bound South for the Equator. BE IT REMEMBERED
THAT the said Vessel and Officers and Crew thereof, have been
inspected and passed on by Ourself and Our Royal Staff AND
BE IT KNOWN: By all ye Sailors, Marines, Land Lubbers and
others who may be honored by his presence that the below
named man having been found worthy to be numbered as one
of our Trusty Shellbacks has been duly initiated into the

Solemn Mysteries of the Ancient Order of the Deep

Ruler of the Raging Main
By His Servant

Commanding
CDR USN

NAME (Last, First, Middle) BARROW, ALSTON M.	Service No. 657 211	Branch and Class USN

Solemn Mysteries of the Ancient Order of the Deep

After the Cuban duty, we again sailed for South America, this time passing through the Panama Canal and heading down the Pacific coast. We cruised close to the coast for our eavesdropping work, and supplemented celestial readings and sun fixes with sonar sounding and surface radar to fix our daily track southward. One night, as officer of the deck on the mid-watch, I was challenged by a Chilean destroyer, which signaled by semaphore for me to stop my engines and permit them to come alongside. What for? Boarding? The captain and executive officer had retired for the night, and while my boatswain's mate rousted them, I ignored their order and split the difference by half.

To not comply with the direct order from a foreign naval vessel at general quarters with its guns trained on me five hundred yards astern, I slyly compromised by cutting my engines to half speed but sailing on. I knew Chile claimed 200 miles into the Pacific as territorial waters and they could justifiably open fire and claim coastal defense against an unidentified foreign military vessel. A few minutes later, the captain reached the bridge and things resolved with no fireworks. I'm still waiting for my thank-you round of drinks.

The next day we docked in Viña del Mar and went to the Chilean Naval Officers Club, where we were hosted by the officers of the ship challenging us the previous night. Our communications officer, an expert Spanish linguist, did the talking. We spent the next twelve hours in fancy nightspots with very fancy food and companionship. Taxis were called up to return us to the ship, and our Chilean officer friends bid farewell and headed home the other way in another line of taxis.

We got back to the States in the spring of 1965, and settled into a slower routine in the shipyard in Norfolk, Virginia. My first diary, the 1965 edition etched in gold on the leather cover with AMB, was bought at the A&F store in New York City. In the front, I entered "Ltjg. A.M. Barrow, USS *Georgetown* (AGTR-2), New York, New York" and on the second page a breakdown of my monthly pay of $656 per month, yearly total $7,872. The 1966 diary bought in San Francisco was saved for the coming year in Vietnam since I knew my tour of duty, if I made it through, was for one year—ending in August 1966.

In the front pages, I carefully spelled out all the names of the seven sailors in each of my four details of deck division on my ship, as I was the first lieutenant or head deck officer of the ship, responsible for its topside—line handling, cleaning, painting, armament, and such. I was also the gunnery officer and assistant navigator. Since we were a surveillance, or spy, ship, we were independent steaming—loners—and carried only one manually operated five-inch/50-caliber gun on the bow to back up the small arms locker holding nothing larger than a couple of 30-caliber Browning machine guns to repel boarders.

Early notes are for deck supplies like four snatch blocks at $30 per fifty gallons, non-skid at $20, cargo nets, and more. I listed names of four stocks with recent prices and stop-loss points, none of which are still listed on exchanges. Those four positions might well have been my first stock investments.

> **JANUARY 5, 1965.** *Underway for West Coast South America from Norfolk Naval Shipyard (NNSY).* [Of New York City, I only remember the trip to Abercrombie & Fitch to buy my pocket diary. The next day was off Hatteras—we made the Virginia coast in the first day's sailing. But diary-keeping started slow.]

> **JANUARY 20.** *Arrived in Rodman, Panama Canal Zone.* [On the next blank page I charted a coastal bay showing an enemy observation post at its terminus and rocky points on the sides—an inlet in Vietnam where we first took fire on a night patrol.]

> **FEBRUARY 7.** *Passed along coast off Lima, Peru, about eighty fishing boats at anchor.* [There are many crossed-off notes in April, lists of things to do while at sea.]

APRIL 1965. Back to North America. We operated out of Key West on surveillance operations in Havana. Every Monday morning we got underway after a weekend on liberty and headed south ninety-two miles to Havana Harbor. We would spend the week a few miles northwest of the port entrance at El Morro castle, steaming in a racetrack pattern about two miles in length while doing our electronic spying. When I had my officer of the deck watch, we monitored and recorded all Russian and Chinese vessels moving through the harbor. Periodically, we were buzzed by Cuban aircraft and we picked up occasional survivors from refugee boats. I had a top secret coded message I could send to Key West Naval Air Station if we were engaged or strafed from air or sea. Whenever we were on station in Havana, we had two armed A4D fighters on the runway sitting in warm-up status. They could be overhead covering us five minutes or so after receiving my message. When Friday finally came, we headed back north the ninety-two miles to Key West and weekend liberty for the crew.

We also spent time at Key West doing maintenance and repairs. I got to know Key West's bars, tourist sites, and eateries intimately from time there during the Cuban crisis. On liberty, we soon found out that the lobstering was productive when we rented a skiff and went skin-diving. One weekend, three of us ventured out to a small coral reef offshore where we speared sixty spiny lobsters. We didn't have licenses, and no doubt many were undersized. Triumphantly, we returned to the ship and delivered them to the stewards in the wardroom. They fixed us a seafood feast of broiled lobsters and yellowtail snapper that night.

I was finishing the seventh of eight years of part-time and full-time active duty, and had gone to sea in a wide range of vessels. The memory of feet on deck for days or months at a time is common to the seafaring experience. Your body unconsciously balances on the deck as you acquire sea legs. In rough seas, you instinctively reach out to brace yourself. Later, when you disembark, you continue the sensation of rolling for hours after alighting on terra firma.

The sounds at sea, depending on type of vessel, stay with you forever. I spent weeks and months on sea on varied types: destroyer, submarine, carrier, landing ship, research ship, wooden diesel junks, and aluminum patrol boats. When a large ship changes speed or course, it may shudder and groan. Bulkheads and overheads creak, thump. Engines rumble, whine, roar, or groan. Below decks, the engine room spaces can be deafeningly noisy, full of clanging, hissing, pinging, and drumming along with the stench of steam, fuel, and hot grease.

> **MAY 7, 1965.** *Make a pre-deployment (South America)*
> *check-off list. Trash is to be dumped over the fantail.*
> *Offload grenades.* In the following days, my diary entries
> indicate we steamed up the Atlantic and were in port
> alongside a tender for repairs and servicing in Norfolk.
> My car was broken into while parked outside the base
> gates. In the next few days, my military life changed as
> Paris and Vietnam intervened. The Pentagon was looking
> for Vietnam volunteers.

ADDENDUM 1
GROWING UP STRADDLING
THE MASON-DIXON LINE

SUCH WERE MY BIFURCATED early years. The original Mason-Dixon was a line separating the southern border of Pennsylvania from southern states. It was drawn by surveyors Charles Mason and Jeremiah Dixon back in colonial times. The purpose was to settle a long-running property dispute, but it came to be popularly known as the boundary between northern free states and southern slave states. In popular culture, it has appeared in numerous songs, movies, and books—even cartoons. I fondly recall Bugs Bunny in *Southern Fried Rabbit*, where Yankee Bugs heads to the greener Dixie looking for better carrots, which is defended by Yosemite Sam who thinks the Civil War is still on. In my early years, the Ohio River at Evansville straddled the line. I spent most of my youth on the Yankee side of the line, waiting for summer when I would cross to the Dixieland side.

When you get old, it turns out that you selectively remember and selectively forget that you remembered things. I can't remember the WWII blackouts in Baton Rouge anymore—I was maybe four years old. But I remember that I recalled experiencing blackouts during WWII when the city lights were extinguished as a precaution against German submarines in the Gulf or the Mississippi River. Remembering events in Louisiana seems easier than remembering events in Indiana, because Louisiana is where I seemed to belong.

My diaries never started until I was out of college. I look back on my early family life as disjointed, awkward, and abnormal. Now, it's obvious to me that many kids in my neighborhood were experiencing the same confusing family life in the 1940s and 1950s as American families, after the Great Depression and world war, struggled to return to normal lives. Did any Chinese, Vietnamese, English, French, or German teenagers consider their growing up years normal after WWII?

Mother and Dad were totally different personalities. The saying goes that opposites attract, and my mother could be seen as introverted, intellectual, sensitive, reclusive, artistic, and a bit snobbish. My father was extroverted, athletic, on the macho side, and was more comfortable outside around male friends engaged in man-talk. The one thing they had in common was a casual disregard for housekeeping and tidiness. Some people put things back. Others leave the toothpaste cap off, the spice jar open. With our style of parenting, I never acquired neat ways, good study habits, or steady self-discipline.

Dad's conversations might be coarse jokes and stories, but I can't recall him ever cussing. Duck hunting, tricks, jokes, adventure were his thing. People who knew him liked him or not—but they always remembered him. He had his ways. He always wore his wristwatch face down with the metal band on top. I finally got the nerve to ask him, "Why do you wear your watch that way? Everybody else wears the face up?" He explained that the glass face was better protected from getting scratched up using his trick.

Mother read lots, had teas with ladies in her social circle, cooked, antique shopped, brooded, nipped on sherry or gin, loved opera and European matters, and dutifully headed for our

Episcopal church every Sunday. We ate fish on Fridays. She was proud of the Barrow family heritage, and became something of a local authority on West Feliciana antebellum history and my dad's ancestors. She certainly knew more about the Barrow ancestors than he did, but he had all the local street-corner scoop.

According to his birth certificate, he was born Alston Pirrie Barrow in Bayou Sara (the commercial waterfront village at the steamboat and ferry landing on the Mississippi), Louisiana, in 1908. It lay just below the hill leading up to the larger town of Saint Francisville. Bayou Sara regularly flooded during high water years when the surrounding low country became swamp all along its course. For days on end, houses, lumber, trees, and dead cows floated downstream toward New Orleans, leaving a backwater stench while the water receded. Over more than a hundred years, the town also endured fires, steamboat explosions, yellow fever epidemics, thievery, and wartime bombardments.

Its charter was revoked in 1926 and the place finally died in the terrible floods of 1927 that left over a half-million homeless people up and down the great river basin. Dad's middle name, Pirrie, died as well. When he was a school boy, he went to the courthouse and had it legally changed to Belhaven. Belhaven was his best friend in school. Pirrie was a boy (likely a cousin) in town with whom he fought. Eliza Pirrie of Oakley Plantation played a central role in Barrow family history as the daughter of James Pirrie and Lucy Alston. She was tutored as a teenager by John James Audubon and later married Robert Hilliard Barrow, the first of seven Barrows to take his name.

Dad had his ways, but I never heard stories of drinking or carousing when he was young. He would have an occasional beer while camping or fishing. If he drank in his earlier years, I reckon he quit drinking after my mother's habit took hold. The genesis of her drinking probably was his going off to WWII, leaving a not-particularly-resourceful wife to take care of two young ones. She had a supporting cast, though; her two brothers and their wives plus her father-in-law and wife lived almost within walking distance of our white clapboard house in east Baton Rouge.

Dad (center) and brothers with wives

The post–WWII generation grew up in the Great Depression, experienced the aftermath of Prohibition and world war, and entered the Cold War period. As adults, many lived faster lives—hard drinkers, heavy smokers, fast spenders living for the present. Eat, drink, and be merry for tomorrow we die, much like the southern rebels in 1864 as defeat closed in. During WWII, large numbers of soldiers were drafted following the Pearl Harbor attack, and even greater numbers willingly volunteered. Dad may have been a draftee. He came home three years later at the end of hostilities to a changed family.

Dad spent much of his youth in the beautiful live oak hill country of West Feliciana parish on his uncle's Tanglewild Plantation, playing and working with Black boys. Home was for school in the cooler months. I suspect that my grandparents were poorer as city dwellers than some of the Barrows who stayed closer to the land after the Civil War. Granddad lived in a modest one-story lopsided house in the old center of Baton Rouge. I knew him as a reserved older gent mainly involved in oil painting and playing classical music on his piano. I never recall seeing him outside his house. Grandmother was blind, having gradually lost her sight from atrophy of the optic nerve.

CAPTAIN
BARROW

Captain Barrow fighter

Getting a loan from his uncle, Douglas Hamilton, at the then-exorbitant 8 percent annual interest rate, Dad went on to Louisiana State University (LSU) just as the Great Depression took hold. He ran track and became the captain of the boxing team, a big NCAA sport in the early twentieth century. As the Depression deepened, he lingered at LSU three more years as the boxing coach. If he had his druthers, he would have studied architecture, but he got two master's degrees—first in geology and then stuck around for another in petroleum engineering. The populist Governor Huey P. Long was pushing LSU and its sports programs toward national recognition in order to embellish his political credentials. Long's aggressive promotion may have been largely responsible for dad's coaching and tuition opportunities. Decades later, Dad was invited back to LSU to the homecoming football game at a packed Tiger Stadium, where he and other LSU head coaches of various sports were honored for their services in a halftime midfield ceremony.

In the midst of the Depression, Dad finally left LSU and headed west in a Ford Model T with DeLesseps Morrison, a buddy at LSU from New Roads, the French town across the river from Saint Francisville. They found work as roughnecks in the Texas Panhandle oilfields. Years later, Morrison become mayor of New Orleans and ambassador to France in the Kennedy administration.

Mother was born in 1909 in Hammond, Louisiana, about sixty miles east of Saint Francisville. Orphaned as a child to her spinster Aunt Florence Hull, I don't recall her ever discussing her childhood. Her father died at the age of 50 in 1911, her older sister two years later, her mother three years later. She grew up deep in Delta country in Clarksdale, Mississippi, where Aunt Florence had a cotton farm. In summers, they journeyed three hundred miles to Monteagle, Tennessee, to avoid the oppressive heat and malaria of the low country. She started at

Mississippi Southern Teachers College and somehow ended up at LSU where Dad studied. Only recently, I found out that she had that older sister, Mary.

I was born in Baton Rouge eighteen months before Pearl Harbor. Early memories are always sparse, but I remember the trauma of getting Tabasco in my eyes. Our maid Dora dragged me out to the driveway, forcing me to stare up at the sun. I cried at the seemingly cruel punishment, but only decades later realized she must have been tearing my eyes to wash away the pepper traces. I loved lying on the rug next to the big wooden RCA radio listening to classical music in the evenings. I recall the night in 1945 when Dad came through the front door in his Army uniform—back from overseas. He plopped his duffel bag on the coffee table, and sat with us on the couch, restarting our family life.

Dad was gone for two or three years, first to North Africa then to Italy via the invasion of Sicily. He took charge of a prison camp for Germans after General Erwin Rommel's retreat. Dad took his jeep out on the North African plains to chase and shoot (with a tripod-mounted machine gun) gazelles for meat to feed the prisoners who complained about brot und wasser (bread and water) meals, mimicking their treatment of Allied prisoners.

Dad had two jeeps, name plates on the front: Beezy (my sister Mary's nickname) then Beezy II. Beezy I was bombed. After Dad died, our family lawyer told me a war story. The German Army near Naples was desperate to retake an Italian fuel tank farm that they had nearly encircled, and the American general told Dad they were going to have to quickly destroy the tank field and retreat. The general asked my dad how to do it safely.

Dad told him to get him an air compressor and give him twelve hours. In the hot midday, Dad pumped the fuel tank voids full of compressed air, and in the chill of that night the entire tank farm's tanks imploded with the fuel all running out as the Germans closed in. After the war, our whole family went to New York City and were hosted to a lavish dinner at the famous 21 Club by the president of Standard Oil of Italy, an executive who spent some time discussing business matters in Italian with still-fluent Dad.

Dad loved photography. He lost his Leica 35-mm camera in Italy, and it was mailed back to him unscathed from Germany after the war, having been registered by serial number. He always had a darkroom in the basement for developing film, and I remember seeing boxes of aerial bombing photos of Italian or German cities.

My favorite early memories are my Uncles John and Francis taking me to their Amite River fish camp where we would spend several days. We would start out in their Plymouth coupe, with me alone in the back seat. I don't recall a radio so maybe they weren't music types. I spent a lot of my early years looking out the open back seat window of cars. After a while traveling through the piney woods of Tangipahoa parish, with maybe a stop at the gas station and a six ounce Grapette pop for me, we would finally slow and turn off through a locked wire gate and ease down a red clay road through the pines. Around a final bend, the woods opened up and the vegetable garden came into view out the left back window—we had arrived!

On the river, we fished and tended to trot lines set for catfish. I helped pick butter beans, cushaw, and Creole tomatoes in the seemingly vast acre-or-so garden tended by Shorty, a Black man with a mule and plow. Uncle John always warned me to watch for rattlesnakes and copperheads in the woods and garden, as well as poison ivy. I was cautious and careful, afraid of dark and shadowy places. Down at the river, there were always slithering water moccasins to watch for. Uncle Francis advised me that snapping turtles eat children's toes.

The wood cabin on a raised foundation was a primitive, creaky, moldy, screened-in shambles built on creosote pilings on the cheap with kerosene lanterns, an ancient wood stove, and a rusting white porcelain ice box on wrought iron legs. The pine floor was warped and I could see weeds through the cracks. It sat high above the river in a dense shaded grove of pine trees, surrounded by mats of pine needles and wild grape vines rising into the limbs. For a five-year-old, it looked too dark and difficult for walking and exploring.

We would stop at an ice house and back the car up to the loading ramp for a big block to take with us. There were no chickens or pigs as there were at Uncle Ruffin's plantation because the camp was closed and boarded at the end of each visit. Sometimes, Shorty took the bus to the camp a day or two early to open the camp and work the garden. I don't know where he slept, but he walked to town with a vegetable basket in the evening.

We caught brim on cane poles, catfish and soft-shelled turtles on the trot line. I loved pulling the line up, because every line breaking the surface might surprise us with something different. The river in those days was pristine, cool with small-mouth bass in the tree shadows and overhangs of bank limbs. Who would ever think there were smallmouth bass in the Deep South? We drifted down the river in his flat boat with a split cane fly rod, throwing hand-painted popping cork bugs. I still have one or two. Uncle Francis caught a water spider and put it in his mouth to show off his fearlessness.

Twenty years later as a bitter old man with nagging back pain, he would rail about the mistakes of President Eisenhower and the Republicans. He complained that he had never gotten his Distinguished Service Cross (DSC). Uncle Francis would explain that a German biplane flew over him on the western front in WWI. He raised his rifle and shot it down just like it was a mallard winging over the bayou. He wanted his DSC another Yank had received for the same feat.

Just recently, my niece found that Uncle Francis served as a second lieutenant in a forward machine-gun nest on the French front in the Battle of Saint-Mihiel led by General John Pershing and in which Generals George Patton and Douglas MacArthur, as well as Captain Harry Truman appeared. He never mentioned his experience on the battlefield to me. He came home to Baton Rouge and married Aunt Marie, a dashing young classmate of Dad at LSU. Aunt Marie's brother was a short, balding, squeaky-voiced vice admiral who didn't like hunting and was a bad golfer. He had graduated first in his 1930 class of 402 midshipmen at the US Naval Academy and had somehow got a DSC as a fighter pilot flying off the USS *Bunker Hill* in the Pacific. Years later, he commanded a carrier division and held senior advisory positions in the Navy and at the United Nations. There's no easy way to earn a DSC—especially if you're sitting almost alone in a machine gun nest in no-mans land.

Mother countered the influence of her sportsmen brothers and their wives, taking me off to be tutored in elementary French and ballet dancing (!) when I was four or five. I still recall fidgeting in my seat during concerts and ballet performances. In summer, Uncle Francis would drive me up to Ambrosia Plantation to spend weeks with Uncle Ruffin and Aunt Belle Hamilton. One traumatic drive up to Ambrosia I pestered Uncle Francis with so many questions that he stopped the car and let me out on the side of the road, driving off. I was still tearlessly stunned when he returned shortly thereafter. Cured of runny mouth, I silently got back in and we drove on.

My stays at Ambrosia were idyllic. The plantation was Hollywood Old South, rows of live oaks dripping with moss, a rambling old house, leisurely days and evenings. In the morning, I got my fishing pole and headed to the pond. For a few summers they had an old white nag, Charley, in the barn. My cousin Sarah and I took turns riding Charley through the rows of trees. At night on the galley or in bed with all windows open and screened, more sights and sounds: locusts and frogs chirping, fireflies in the shadows. The cool country southern smell in the air at night and moonlight through the live oaks remains with me.

Some mornings I went with Uncle Ruffin to feed the chickens as Black boys were walking through the foggy pasture to the well with their empty buckets. Back then, they lived in nearby sharecropper cabins with no running water. We went to the grain bin to get a bucket of feed corn, then to the coop. We shucked the ears right onto the dirt as the chickens scrambled and pecked around our feet.

Uncle Ruffin always started shucking the end of the ear then handed it to me; I had trouble pushing the kernels off with my small fingers. I collected eggs with my basket, always peering into the shadows of the hen house for snakes or rats. When it was killing time, I shined. Nimbler and less threatening than Uncle Ruffin, I caught whichever condemned chicken he pointed out. I would hold it by its feet on a stump and then Uncle Ruffin came over and chopped off the head. It would flop away, spewing blood from its neck. That was a chore way more traumatic than the feather plucking to come.

My playmate was Wash, a sarcastic little Black boy who walked barefooted to the plantation house from the nearby cabin where he lived. Youngest in a big family, he was yelled at to wash up before coming inside the house so often that Wash became his name. Wash and I would ride in the bed of Uncle Ruffin's pickup when he went to shop in Saint Francisville or out to Rosedown Plantation. At Rosedown, we helped Uncle Ruffin's spinster aunts maintain the slowly fading but still splendid grounds. I can't remember whether Wash attended school in the fall. Later, when I was in college, I asked after Wash and was told he was "up the road" in Angola state prison, the home of Lead Belly's music and annual prison rodeos.

Most evenings at Ambrosia, after supper, we would go out to the rocking chairs on the front gallery. I swung in a hammock listening to the locusts, watching the moon and clouds up in the oaks, and following the adult gossip. I remember one night they argued on and on about the merits and defects of a relative I didn't know. When I asked about him, Aunt Belle advised me, "Oh, you never met him. He died in the Civil War."

What sense of family and belonging to a place did I absorb as a ten-year-old? When I stayed in Saint Francisville I learned that it was the old home of my family. Later, I learned that the nearby plantations were often built by Barrows, and around the parish many spots were connected to my ancestors.

As for the Barrow family, I learned much later how the War, Reconstruction, and changing times altered the family's fortunes. From antebellum plantations scattered throughout Louisiana, family owners retained ownership of only a few remaining homesteads, little of the land, and only one home under unbroken family ownership—Rosale Plantation in the Robert Hilliard Barrow family. I never knew my paternal grandfather worked with mules on the river levee and scraped out a living at the oil refinery in Baton Rouge. Family members never talked about their reduced circumstances in my presence.

When I was seven, we moved to Evansville, Indiana. Dad first worked for Superior Oil Company in the tri-state area of

Illinois, Indiana, and Kentucky. This was a shallow limestone geology primarily drilled for stripper wells with short-life yields of a few dozen or less barrels of crude output per day. In 1939, he discovered the richest oil field in that area. Upon receiving what he saw as a miserly Christmas bonus of $500, he disgustedly quit on the spot—vowing to never again work as a payroll employee.

He restarted in 1940 as a self-employed geologist. Evansville fronted the Ohio River just east of where the Wabash River, separating Illinois and Indiana, dumped into it. Coming from Louisiana, the land was bland rolling farmland with little mystery to me. Oil traded at only $3.50 a barrel, but Dad could pick a site and drill an exploratory well down a few hundred to a few thousand feet on a $5,000 outlay. His percentage of developed well paydays versus dry hole hits was high enough to deliver a bumpy, but sustainable, income. He could front dollars or prospect and drill by taking a share of output.

A producing stripper well might deliver ten barrels a day for five years, and Dad's share might be an eighth of output. Half the holes might be dry and plugged, and the other half were slowly depleting producers. He sited and drilled wells as fast as possible to build what amounted to an ever-changing stream of depleting annuities, each well generating income of a few thousand dollars per year per well. He spent lots of time away from home. Those income streams were unsustainable and unpredictable, but dependably produced surprises—good and bad. He would come home in a blue funk when the latest favorite well near Mount Vernon suddenly went dry.

I attended Harper Elementary School, a public school about a mile from home on Ruston Avenue. I walked the mile to school and back each day, often running partway down the sidewalk. By sixth grade, we had moved to a much nicer house two blocks from school, but I was transferred to a new public school named Dexter. It was further away, but had no bus service near me. I still walked back and forth to school. One morning, after I crossed busy Lincoln Avenue, I heard screeching tires behind me and turned to see my little family terrier, having trailed me from home, lay dead in the bloody street. If there were students riding buses to school back then, I don't remember them. I yearned to move back to Louisiana and the familiar landscapes, relatives, Spanish moss, biting fish, gossipy girls, southern drawls, and mentor uncles.

Approaching my twenties, I had too often been bored and impatient with friends satisfied enough with hometown life as both destination and security blanket. For them, it was enough that the known world and family would endure after they were gone. I loved Louisiana whenever we returned; I was often bored and frustrated by having to be schooled in Indiana.

I now realize that my older sister and I endured a broken and sometimes traumatic home life in Indiana. Certainly, there were good times, but ours was not a normal life. It scarred us but made us stronger and more resilient. We grew up in an unbroken, broken, healed, re-broken, and healed again home. (My parents divorced and later remarried.) Embarrassment, loneliness, frustration, anger, and resignation were all at work, and we were both relieved to be off to college when the time came.

In summer, I got to spend two weeks at Camp Carson, a Boy Scout summer camp near Princeton, Indiana. I recall initially feeling that I was being shipped off to lighten the parenting load at home. Eventually, I adjusted and stopped getting homesick. I enjoyed nature courses and shooting at the rifle and archery ranges. Mysterious sex jokes were told over the campfire; we didn't quite understand but everyone doffed their caps and slapped their knees.

The second summer was a big year for me. In the fall, my dad and I were called downtown where I was presented a gold medal and engraved shooting trophy for being the outstanding marksman at camp that summer. I had shot a .22 rifle for near perfect scores at the range. I was ten years old, and the thrill of public recognition comes back to me when I look at the award presentation photo which made the front page of the *Evansville Courier*. I don't remember anyone congratulating me or even commenting on seeing my picture.

Receiving shooting trophy and gold medal

A few events from early school years stuck with me. Like my first date in the seventh grade. Mother drove six of us to pick up our dates and deliver them home, corsages and all. I remember waiting an eternity for her in a tiny, cramped living room, but don't recall where we went. Taking her home, I was in the middle of the back seat and when my date clambered out to go to her front steps, I stayed in the car. Mother later scolded me and explained the neglected escort's protocol the next morning. A half century later at a high school reunion I apologized to her—remorseful all these years. She raised her eyebrows saying, "Mac, I don't even remember you on that date." Truth or retribution, I'll never know.

One spring vacation when business was good, we drove down Highway 41 through Tennessee to Anna Maria Island, Florida, where we rented a lone beach house for the week. Dad and I rented a boat and fished in the Suwannee River on the way. I was back in the beautiful South with its special smells and mystery, Spanish moss and blossoming orange trees. The southern live oaks, the warm weather, and the crystal-clear water excited me plenty. Another southern paradise! Later, we toured Silver Springs, peering down through the glass-bottom boats. I pointed out every largemouth bass swimming into view. I resolved to eat seafood every meal we ate out in Florida until the end of our trip, and did.

Brother John and sister Anne came along in the late 1940's. When they were still very young, Dad had a terrible accident driving one foggy morning on the highway north of town. A farmer entered the highway from a gravel side road without stopping; Dad hit him broadside. Even though he was only going 45 mph, he had no chance to swerve. The farmer was killed instantly. Dad's wood-paneled Ford station wagon accordioned, with the wood exploding into splinters and the front bumper compressed to within 10 feet of the back one.

He spent two months in the hospital with a back fracture, broken ribs, a concussion, a broken limb, and more than 200 stitches in his right arm alone. After discharge, he was plagued with headaches and sore muscles initially, and later with itching scars (with an additional diagnosis of adult-onset diabetes). He scratched his back on door jambs like a horse on the barn corner, and the bathroom stank from Absorbine horse liniment. Those months were challenging for us.

The other memorable vacation was a summer trip out West. It was fun but I sensed we were being taken out of town while Mother was in treatment somewhere. It was only Dad, sister Mary, John, and me. Miles and miles of slow travel on two-lane highways and gravel back roads in a station wagon towing our luggage trailer. No A/C so we drove with all windows open blowing a hot breeze, canvas water satchel suspended from the front hood ornament. Sometimes we picked up a radio station with hit songs as we drove relentlessly on. We wiggled and fought and napped in the back seat when not staring out the window for something new.

I think it was the typical 1950s family car tour: Aspen in summer, going up a ski left with no snow for the view, the Black Hills, Mount Rushmore, the Rockies, the Indian trading posts. We camped in Yellowstone in our smelly Army surplus wall tent and Army blankets. In those days, black bears lined the sides of roads waiting for food handouts from cars. Dad had his loaded WWII Colt 45 under his cot, but nightly black bear invasions of the campground and a freak July snow storm finally forced us to pack up. The unseasonable cold front dropped a blanket of white snow, and we left shivering the next morning.

Two days later, we stood on the rim of the Grand Canyon in 106°F heat surrounded by sightseers. On to Carlsbad Caverns and dark, dank cool mystery. Heading back east through the Texas Panhandle, we pulled in to the small town of Flatonia, where our one-wheel handmade plywood luggage trailer had a flat. We waited for a replacement Cessna aircraft tire for two days. Flatonia's only attraction was a landscape littered with petrified logs and wood. Next, we visited "rattlesnake hill" near a drilling rig site where he worked in the depression. We climbed around in the hot sun, but found no trace of reptiles. On to gin-clear Devil's River, where he sunned his back scars and got sunstroke.

My first experiences with grown men were mostly with my mentoring uncles since Dad was off in the Army during the war. After the war, life revolved more around him at home—hunting, fishing, travel, his stories and jokes. When he was young, he spent a lot of time at his Uncle Douglas's Tanglewild Plantation, north of Saint Francisville near Angola state prison. In the 1920s, Tanglewild was probably one of the largest plantations in the South, stretching for thousands of acres. Douglas was a portly, taciturn bachelor, a maverick loner who liked me and was quick to banter and tease.

Dad worked summers there for wages. One summer he and another boy were sent up the road to Woodville, Mississippi, driving a string of mules to market there. Dad told me that he and his buddy, mounted on horses and carrying whips, had a tough trip, taking two days to go the fifteen miles north as unruly mules kept breaking away from the string or refusing to move ahead or go to water. Many years later, Uncle Douglas asked Dad to find him a new pickup truck. I think Dad felt obligated to help his uncle with the chore but worried who would pay what. Something grated on Dad. I found out that Uncle Douglas, decades earlier, loaned Dad money at 6 percent interest for tuition at LSU—what seems like an exorbitant rate for the late 1920s.

After Dad's death, I saw a picture of Dad and his brothers at Christmas time, glasses in hand and waving to the camera. Older brother Montrose, grinning, was waving a four-fingered hand. A relative asked me if I knew how he lost his finger. He told me that Dad was chopping firewood one day when they were teenagers, and Montrose told Dad to stop chopping because he was not doing it right. Brushing him off, Dad kept chopping and Montrose reached down and grabbed the wood. Dad told him to remove his hand or he would chop his finger off. Montrose refused and dared him—and Dad chopped his finger off! Much later I found out it was simply an inside joke of theirs—Montrose lost his finger in a job accident.

Commenting about the two in his brief *My Autobiography to my Sons,* Granddad said "My oldest boy, Montrose, when growing up was rather of a quiet disposition, liked to read and did not care much for outdoor life, always a serious type and studious. Alston was just the opposite, a rough and tumble type, loved athletics, outdoor recreation, loved to box, and was always a leader in any group. I recall him once being in a fight downtown, he was with another boy and somehow got into an argument with some older men. They proceeded to jump on the other boy and Alston waded into them too and then it was a free for all, but Alston came out victorious, with the men sprawling on the sidewalk. They were all taken to court and the Judge had to laugh, as Alston was then a lightweight and the men were large and heavy set, and they showed the scars of battle and Alston didn't have a scratch as he was a good boxer and knew how to handle himself."

On to Bosse High School, about three miles away. I mostly walked to and from school. When I became old enough to drive, I still mostly walked unless friends with cars gave me a ride home. I don't know why I didn't buy a car. Two of my four best friends had a car. I had a savings account holding my earnings from work every summer. When I was nine, I began sacking groceries for tips at Wesselman's Finer Foods, and went on to pump gas, unload rail cars, work on the river, and help out at the local sports stadium.

As for school, my grades were good but not great; I was not a disciplined studier. Dad had me taking German, the language of science, but Spanish would have been more useful. I enjoyed history and literature. Even before high school, I read and maintained stacks of *Field & Stream, Outdoor Life,* and *Sports*

Afield magazines in my bedroom. I read the entire kids' collection of famous Americans—Washington, Benjamin Franklin, Jefferson, and Lincoln but especially enjoyed explorers like Daniel Boone, Davy Crockett, and Kit Carson.

I was sold on Teddy Roosevelt's turn-of-the-century romantic idea of the true man, a rugged individualist comfortable chatting in the parlor, the theater, or over a campfire after a long day hunting big game on the prairie. In school, my favorite subjects were literature and history classes.

One summer, Dad and I put in a line of post holes for a new fence running behind the house. I had been reading Frank Lloyd Wright and mentioned that Wright successfully tested his theory that a column with the smaller end placed at the bottom is stronger than the traditional big end at bottom. Put the posts in upside down? This applied to concrete-building architecture, but I extrapolated to suggest that we try putting the posts in upside down. Astounded, Dad stopped, dropping the posthole digger. "What did you say?" Apoplexy! He was without words; his son was challenging millennia of history of fence construction. I soon let pass my architectural theorizing.

Every summer when I didn't have a job away from home, I worked around the house. One year I spent half the summer scraping flaking white paint and repainting our four-bedroom lap sided house. I hated every hour in the boiling sun. The teenager pay rate was about forty cents an hour at the time, but it all added up. On the spur of the moment, I spent $105 of my savings to buy a new 12-gauge Model 12 pump shotgun. Otherwise, I could have afforded a Ford, Chevy, or Plymouth sedan off a used car lot like my friends Bob and Kenny. Thankfully, they would come by to pick me up many nights. Later, after getting my license, I would borrow Mother's garish coral and white Lincoln sedan. On nights with neither, I was at home, grounded.

Of course, in those days a cheeseburger, fries, and coke were forty-five cents. Gas was thirty cents a gallon. For my entire adult life, I have endured sporadic dreams about looking for a ride home or walking strange side streets from school to home in my embarrassment at having to walk home as an upperclassman in high school. Sometimes, classmates would stop and give me a left. Just as often, they would ride on by waving and maybe laughing.

As I said, I tolerated growing up in Indiana, but my birthplace and sentimental home was Louisiana. Decades later at my sixtieth reunion, I recall a friend and I surviving the rubber-chicken banquet for our shrunken class, then driving around a modernized town mumbling to each other as we drove through familiar neighborhoods. Boy am I glad I left this place! Whew, I'm glad I got out of town after graduating! All the classmates who never lived out their lives more than one hundred miles down a cornfield highway, the ones we thought might be novelists, senators, MLB players, Broadway stars, or even president. Well, we weren't famous either, but we saw the world and made our fortunes and did it our way.

My great love was fishing. I never had much luck in Indiana, in Yankee land. Go south for fish. I recall Mother dropped me off for an afternoon at a public lake on the way to Newburgh, up the river from Evansville. I was the only fisherman on the boat dock and soon found out why. Mid-summer swimmers splashed in the café con leche water at the nearby gray sand beach. I dropped my bait, probably a dough ball, into water that swallowed it in the first couple of inches. Stunted bluegills the length of my pinky came up on my hook every cast. I was astounded: Wasn't there even one hand-size fish in the lake to trim the hoard of these pitiful creatures? I had never seen a pan-fish so small in Louisiana.

Dad and I made a couple of spring trips down to Kentucky Lake. The first trip, he had a buddy take us to brush piles along the lake's edge. Our Missouri minnow baits yielded schooling crappies right away, often two at a time. High fives! Finally, fun fishing! The next spring we tried it again, alone in our own skiff. Dad wasn't much of a fisherman and it showed. We couldn't locate the right brush piles, skulking home skunked. That was our last trip to Kentucky Lake. A decade later he purchased a small piece of land on the lake, keeping a trailer and boat there. I never visited the property—had the fishing bug bitten him? Had he found a buddy guide to educate him on fishing tech-niques?

There was a special workaround for my fishing mania. During high school, we would drive up to Ontario in the fall for a week of duck hunting and fishing for pike, walleye, lake trout, and smallmouths. One of his high-roller drilling buddies—let's call him Tex—would drive us across the Wabash River and up the length of Illinois, mostly at 90–100 mph in his big Cadillac. We headed to the north border and Ontario, making this drive three years in a row. Invariably, I was in the big back seat watching the corn fields fly by, and I hear a siren. A burly Illinois patrol-man on his Harley pulled us over, greeted Tex (and me?), and checked his license. A brief low conversation ensued. Tex pulled a $100 bill from his wallet, handed it over, and we headed back out onto the asphalt bound for our Canadian fishing adventure. A few miles later, the speedometer read 100 again. I remem-ber Tex's alligator-skin wallet, twice as thick as Dad's. Tex had hunted tigers in India on an elephant, Alaskan big game, African lions, you name it. All his trophy heads stared at you when you pulled into his three-car garage. The one "game" he couldn't dominate, his wife, moved the heads from house to garage walls during a long absence on safari.

On a special side trip we drove up to Nipigon, Ontario, and hired an Ojibwan and his canoes to paddle and portage us in to a remote lake. Maybe fifty acres, the crystal-clear blue water deep in the silent north woods held mainly two creatures: brook trout and freshwater shrimp. We set up my Dad's musty canvas wall tent, camped, and caught bright salmon-orange fleshed brookies by the dozens. I found an inlet stream and proudly took two-pound brookies at its mouth with my first hand-tied trout streamers.

With Tex and fish haul for Obijwan tribe
the North Woods Canada

Another Obijwan took us out in canoes to a wild rice lake. The stalks were heavy and bent with grains ripe for harvesting. It was clear and cold and the ducks arrived as the evening sky turned golden. We heard them far off at first, then watched clouds wing toward us over the trees, swarming from every direction.

In moments, we were covered in ducks circling and alighting on the water. Instinctively, I ducked and could have hit teal, widgeons, bluebills, mallards, shovelers, canvasbacks in the air, like mosquitoes, with my paddle. We swung our shotguns to fire in front of any ducks gliding down further away than ten yards or so. As the sun disappeared, ducks were still landing in the rice all around the boat as we paddled and scooped up the floating bodies. Hundreds more arrived at dusk, determined to land as we left the lake guided by our flashlights.

My first downed game was a goose at the age of nine. We hunted in the winter for geese at Island 21 in the Mississippi River off Cairo, Illinois, where Dad was a charter member of a twenty-member hunting club. We hunted geese there each winter, and I became a good shot by my teens. I remember walking across a sandbar with Dad's friend Ray when a pod of three Canada geese flew forty yards above us as we hunkered down. Ray said shoot, and I shot once with my single-shot 12 gauge, reloaded and shot again, reloaded and shot again, all in about three seconds (holding two shells between fingers at the forestock). All three geese fell out of the sky, and Ray was furious, shouting, "Every time I got ready to pull the trigger, you beat me to him." The limit was two per day, so Ray got one, me three.

Ray and I camped out for a week over Christmas holiday in Dad's wall tent. We had dead geese hanging from every tree limb near the tent, and an iron cauldron of goose breasts simmering all week. We would add beans, carrots, potatoes and more each morning. Ray had a whiskey quart to keep him warm in the tent. He was missing two fingers, which had gotten caught in a drill bit coming out the hole. Ray liked me because he knew I had grown up sleeping many winter nights in the oil fields in the back of Dad's station wagon between cases of dynamite, prima cord, smelly drill bits, and core samples. I learned to sniff a core sample to see if it held a show of oil, and I knew a worn-out diamond-studded drill bit from one still usable, and where Halliburton seismograph teams worked better than Schlumberger's.

Winter goose hunting started with a long drive from Evansville across the Wabash River and southern Illinois in the middle of the night to a motel on the Mississippi River near Island 21 Hunt Club. At the time, the club owned half the island and the Horseshoe Lake Federal Game Preserve owned the rest. The motel lot was jammed with cars and trucks. Upon entering the front door of the restaurant, you were blasted by the music and talking crowd of hunters, drinkers, and floozies at the bar. We would go through two more doors, past gamblers at slot machines and card tables, and into the dining areas with enormous slab tables and benches.

Starting at 4 a.m., waitresses would begin loading the tables with coffee pots and platters. Breakfast was one price all-you-can-eat. The first two-foot platter sliding down the table might hold three dozen fried eggs, the next one five pounds of bacon, the next one Polish and patty sausages, the next biscuits and butter sticks. I watched in awe as big lumbering men, some in hunting clothes, some drunk, would scrape five or six eggs onto their plate, add a pile of bacon, and finish it all in a couple of minutes. This was a loud, rough crowd with whiskey pints, cigars, and beers next to their coffee. After breakfast, we left the hot, noisy building and went to our car. We got our guns and gear and headed out into the frigid dark, under a roof of stars for our goose pits out on the sandbar by the river. It was a half-hour walk in the dark, usually in bitter pre-dawn cold through mud, ice, and frozen fields.

Hunting was from pits out on the sandbar by the river. The pits were about ten feet long and five-to-six feet deep, lined with a bench and plywood sides. The sliding cover was two wood frames covered with burlap and porch screen. Already hearing distant honking in the lighting eastern sky, we hurried to set out our decoys up wind and within range of our 12-gauge no. 2 shot with one-and-three-quarter ounces of lead pellets. In the blind under our screen cover, we waited for shooting hour, trying to stay warm as we sat still and quiet at daybreak. My dad would light his little Primus white gas camp stove and we took turns holding our gloves over the flame. The honking grew louder as a string of geese approached, and we eased the cover open on each end and honked back. When they turned in to our decoys, they dropped lower—heading straight at them. Or a few might peel off from a larger flock heading up river. The thrill of facing these big birds, wings set gliding straight at you going 50 mph, wings set, fifty, forty, thirty yards—finally able to see their glistening black eyes. Then we'd hear the whoosh of their wings, and we jumped up, stock at shoulder, picked a bird and started shooting. In those days, the limit was two geese but sometimes we would take a mix of mallards and geese.

When I got my first goose, I shot a Stevens single-shot 12-gauge hammer-cocking model. I held my second and third shells between my fingers under the forearm, breaking and ejecting and slamming the next shell in the chamber, then closing the breech and cocking the hammer, aiming and firing all in a couple of seconds. At my size, the recoil kicked me back nearly a foot.

The sub-freezing temperatures and wind out on the river bank always tested me.

Children's gloves and boots seemed never adequate. My dad got me a white gas Jon-e Hand Warmer but it kept going out. One day, my hands and feet were so frozen that Dad gave me a book of matches and pointed at the woods across the sandbar a quarter mile away. I got out of the blind and started walking. In the woods, I collected twigs and sticks in a pile, but my fingers were so frozen that I couldn't hold a match to light a fire. I managed to warm up by running back to the blind.

The winter I was ten or eleven, I was walking along a cut-through sandbar in the woods when a pair of geese rounded the corner of the woods headed straight for me. Crouching down, I rose and shot, downing one of the two, but the other flew on. The next day, crossing the same bar alone again lost in thought, I looked up to see a lone goose headed straight for me fifty yards out at head height. Instinctively I ducked down on my knees and raised my gun off balance to shoot as it nearly hit my head with one wing. Was it the mate from yesterday? Was it trying to knock me down? Yes, I'm sure that was the mate. I never pulled the trigger.

That experience, trying to understand the death of a loved mate, made me regret my earlier shot. I had been proud of my shooting prowess. There was a moment of reflection where I suddenly had an epiphany, grasping the meaning of death. Whenever I died, it would be the end of everything. I would no longer exist. I would never go fishing or eat bananas again. I wouldn't be me! Much later, in my seventies, I managed to reach an understanding. When you die, your identity leaves, so your being surely disappears. Since you no longer exist, then by definition there can no longer exist any sense of loss or remorse except in the memories of others who knew you.

Death's horror dies with death. On the bright side, some wit had the idea that heaven's gate may well bring you to a reunion with every dog in your life that loved you wagging his tail in the prime of his life waiting for fun and feeding. I like to think of it as *fidopiscicism*, the fun religious order.

Hunters learn that animals are like people. Quail have little brains like some small game. They flush instinctively, and then figure out which way to go. Leopards and hyenas connive and skulk then mercilessly attack. Elephants with bigger brains seem to ponder before deciding, but sometimes panic, too.

Most summers when I wasn't working, we drove back down to Louisiana where I would stay at my Uncle Francis's house in Baton Rouge or at Uncle Ruffin and Aunt Belle's Ambrosia Plantation near Saint Francisville. For fishing, Uncles Francis and John jointly owned a forty-eight-foot twin engine ex-Coast Guard mahogany-hulled cruiser tied up in Cocodrie at the bitter end of the gravel road south of Houma. This was the heart of Cajun country. We would journey all day from Baton Rouge, timing our trip to catch the ferry across the Mississippi. Arriving in the evening, we would board the boat, stow our three hundred pounds of ice in the ice boxes on board, and head over to a Cajun café for a supper of fresh boiled shrimp and salad. The small shell-on culls fresh off the shrimp boat boiled in Cajun seasoning were the best shrimp I have ever eaten. Where were the big shrimp I saw in the nets and boxes? They laughed and said they all went to New York.

I don't think the boat had a name, but I loved it. Varnished mahogany deck, bulkheads, and overhead, a screened dining compartment, plus a front hatch in the forward bunk room for climbing in and out to set the anchor or handle bow lines. As we had no air conditioning, my uncles simply wore pajamas at sea with the legs cut off at the knees. Flank speed with the twin Chrysler diesels was about thirteen knots—it took a few hours to go down the bayou and out to Wine Island where we would anchor for the night. You smelled oil, diesel fumes, marsh grass, fish, and salt breeze all mixed together. Shrimp boats and converted PT boat crew boats of the drilling companies threw up waves when passing us going in and out at twenty-five knots.

Shorty, the field hand from Amite, hugged the rail in his kapok life vest. He couldn't swim but loved the fishing and the boat.

At Wine Island, we took the skiff to the beach and cast for speckled trout while wading in two or three feet of water, throwing Mr. Champ spoons with a bucktail or a plastic-molded Fishermen's Favorite lure. When hooked up, we would back up and drag the trout onto the beach. The trout were usually small, but some days we caught more than a hundred. Other days we fished for redfish in the bayous, or journeyed offshore to the rigs to catch lemon fish (cobia), bluefish, and mackerel.

Uncle Francis would put me alone in the skiff with a hand line and spadefish for bait, and I would paddle over to the rig and slap the water with my paddle. A pod of lemon fish would invariably swim over to investigate, I would toss my baited line out, and the quickest fish would take my bait. These fish averaged perhaps fifty pounds, so I would jerk to set the hook and simply brace my feet and hang on as it towed the skiff around until it wore down. Back at the big boat, I tied up and after climbing aboard Uncle Francis would take my hand line and shoot the cobia in the head with his pistol before hauling it aboard. They made delicious fish steaks.

My favorite was night fishing with lantern and spear for flounders in a foot of water on dark calm nights at high tide. Sometimes we came back before midnight with more than fifty flounders. The real fun was seeing nightlife in the illuminated shore water—iridescent squid, crabs, stingrays, sand sharks, starfish, mullet, and schools of baitfish—in the surf feeding or fleeing predators. One time I saw an eight-foot alligator gar in the surf. Another time nearly every square foot of the shallows at Wine Islands was covered in mating blue crabs—millions of crabs. Wine Island, which had a resort hotel in the 1880s, is gone, sinking beneath the Gulf waters into a shallow shelf in the 1980s.

I was a little older, maybe fourteen or fifteen, when I took my last fishing trip out to Wine Island on my uncles' boat. We tried to go floundering the first night, but the water was too rough. The next day we fished for reds in the bayou to stay out of the wind. Back at Wine Island that night, the storm hit after supper. Winds rose from the south to over 50 mph and we put out a second anchor after moving into the lee of the island, running both engines at idle into the wind. By midnight, we were running both engines flank speed—all of twelve knots—and still slipping away from the shallows. I went below to sleep after tiring of bracing myself against the violent rolls. I got seasick right away. Back topside into the wind and rain holding on to puke over the side on downswings. The next morning, the wind died. Our fifteen-foot wood skiff, an old friend secured from the transom with two one-inch manila lines, was gone. And the outboard engine, too.

Under breezy gray skies we searched around the island and found only one oar and one orange life vest. I tried fishing while we prepared to go home, and the only thing biting in the muddy waters were catfish. My uncles were shaken by the ordeal and I knew that we had made a narrow escape. We eased up the bayou and into our dock slip at Cocodrie.

Back in Evansville, I finished high school looking forward to leaving the Midwest for places more exciting, intent on fashioning a more adventurous life than schoolmates just drifting through it. Louisiana was where I had always yearned to be, but my love of literature had exposed me to the adventure ahead—if I could just extricate myself from the present. I had already finished most of Hemingway's novels and short story collections by my senior year, along with many of the classic adventures in books by Mark Twain, Robert Louis Stevenson, Jack London, Rudyard Kipling, and other writers available in the library.

Dad wanted to get our boat, an eighteen-foot Lyman lapstrake inboard with a forty horsepower Gray Marine engine, down to Saint Francisville where it would be useful for duck hunting and catfish trotlining on the Mississippi. Dad was getting older and his drilling business in the Midwest was winding down. He did more contract work for the Mulzer brothers well drillers and stayed around home more. After graduation, he said, "Get a couple of buddies and you guys can take it down river. Go down the Ohio to Cairo, take a left at the Mississippi down to Bayou Sara, tie it up, walk up the hill, and call the house from the post office and someone will come pick you up." Dad had worked on the river ferry as a kid. I wonder if he wanted to go, too.

Getting Dad's boat ready to head down
the Ohio and Mississippi Rivers

I got Bob, who was a best friend, and another classmate to go with me. Bob was fascinated with Dad's stories and jokes and his chancy work. I suspected Bob's dad, a contractor, was a steadfast, predictable home-bound type—the antithesis of mine. Dad liked having Bob around.

We launched the boat at the Evansville boat club in early June, a few days after graduation. Lots of advice on safety, navigation, camping, and map reading. Dad got us a set of river pilot charts, a huge book of three-foot-by-eighteen-inches charts showing page by page the Mississippi River channel, ten miles or so per page with depths, bars, islands, obstructions, locks, lights, landmarks, and towns. It was our bible for the next two weeks. I had a copy of Mark Twain's *Life on the Mississippi* to read while traveling. I remember we set out down the Ohio in high spirits, excited to be off, loaded with provisions, extra gas cans, and camping gear. We made it the two hundred miles down to the junction of the Mississippi without incident.

It's funny, I can't remember many details of the trip—it's been sixty years! I recall we had to navigate the locks, learn the protocol for getting permission, waiting our turn, entering the lock, and so forth. I remember passing up mud bars looking for a good camping spot at the end of the day. I remember trying, frequently without success, to catch catfish. We steered around bends, avoiding barges, and stopped at landmark places like Memphis and Vicksburg.

Somewhere below Memphis, off Coahoma County, our manual fuel pump broke and we drifted downriver for two days, staying out of the main channel to avoid barge traffic, until we arrived at the shore of Greenville, Mississippi. (I knew Coahoma County well as mostly cotton fields and mud. The county seat Clarksdale was the home of many blues singers and writers. If you ever were there more than a month or two, you might write blues songs, too!) It was where my orphaned mother was raised. Years after this trip, I owned a family cotton farm there with my siblings, a family inheritance.

It was a long slog through mud and up a cliff to find our way into town, where we ordered a replacement fuel pump at the hardware store. I think we waited two or three more days until it arrived. We located beer and hot dogs and hung out, putting out the trotline for catfish. I don't think we even had a portable radio so we probably just sat around talking, whittling wood, reading, and slapping mosquitoes.

Finally replacing the old manual fuel pump with an electric one, we got underway motoring down past Vicksburg and Natchez and into Louisiana home country. We pulled into Bayou Sara one morning, tying up near the free ferry landing and climbing the hill into Saint Francisville to use the pay phone and tell my family that we had arrived. We achieved our objective—we felt a sense of accomplishment, plus relief from my Dad that his boat and its occupants had made it. That fall, I went to college and never saw the boat again. Bob took a Greyhound back to Indiana and I next saw him twenty-five years later at our high school reunion.

Later in the summer, I joined a large group of college boys at Southwestern Publishing Company's auditorium in Nashville to learn the ropes of selling their family bibles door-to-door. I teamed up with Harry, a fastidious concert pianist and Vandy upperclassman. We headed out to Williamsport, Pennsylvania— our assigned district for sales. We found a dumpy little apartment on the south side, settled in, and headed out on foot in different directions to knock on doors. Harry was older and experienced at selling, so he did better.

We worked separately, and I usually saw him only at the end of the day. As he was a bit of a stick-in-the-mud that was fine with me. I didn't get any selling tips. This adventure was hitchhiking all over the county and knocking on doors where recluses, religious nuts, and others lived. After a few weeks of canvassing our territory by thumb and foot, we took leave while waiting for our bible shipment to arrive for hand delivery to our purchasers, who we required to put down a deposit. We would have to take each bible to the home and collect the final payment. We headed out of town by thumb for the big city, making it to Greenwich Village before nightfall with a lucky ride. It was my first time in New York City and Larry headed us straight to Greenwich Village, the fabled home of beatniks, artists, musicians, con men, and perverts. I was fascinated with the scene.

At some point I lost Harry, or he lost me. I managed to spend my money down to less than a dollar, took a subway uptown, and began walking out of Manhattan across the George Washington Bridge. Over in Jersey, I got a ride to Scranton, Pennsylvania, where I was dropped off at midnight. I was then picked up by a pervert, got him to stop the car, and I jumped out. I walked the twenty miles to Wilkes-Barre in the night, arriving near dawn on the outskirts where I found a bike on the lawn of a house. Jumping on that bike, I coasted downhill in an exhilarating cool breeze, the best bike ride of my life. I left the bike at the front door of the police station and continued walking, finally thumbing a last ride into Williamsport and falling into bed back home at my empty apartment. I awoke a full day later.

A few weeks later, I got wanderlust again, heading by thumb up to Toronto. My primary memory of the trip was getting a ride at night in Batavia, New York, with a fat, drunk lady holding a beer bottle between her legs driving 80 mph in the night complaining about having lost every race at Batavia Downs. Alarmed, I demanded that she stop the car and let me out.

In Toronto, I made sure to keep enough money to be able to buy food and drink on the way home. I took the easy way out going partway home on a Greyhound. The Finger Lakes region is a beautiful area of the country, and passing by Niagara Falls was memorable, but I recall little of my short visit in Canada. Years later, while working on Wall Street, I made several trips to bank and insurance company accounts based in Toronto, leaving me with a revised recollection of a city seen before from the bottom. I may have been a poor tourist in those days, since my main objective was simply to get somewhere. However, touring a city on foot with a backpack and only a few bucks prioritizes finding a place to sleep and something cheap to eat.

I spent the rest of the summer in Saint Francisville before heading off to college. Uncle Francis had a green Plymouth, an old four-door with fins he let me drive. I spent a lot of time dating in Baton Rouge and hanging out at the country club. My Aunt Marie got me a blind date with Peggy, a pretty black-haired high school senior living with her parents just off LSU's rolling campus golf course. We wasted away the summer at movies, parties, and whatever else the town offered; in August, we sadly parted. When I came home for the holidays as a Vandy freshman and tried to call her, I was too late and too long gone. She had moved on. For me, I realized my growing up days were over. Now it was time for college, the Navy, and to see the world.

ADDENDUM 2

AS I WAS FINISHING this memoir, the following life story by my great aunt came to me from family papers in Louisiana. It's been only lightly edited to retain the original sense of the story.

When I was a boy, my father used to take me to visit "Meme" as my great aunt Hazel was nicknamed, along with visits to my other great aunt—her big sister—my Aunt Isabel who stood six feet two inches tall. Her story is a very enlightening window into the postbellum life of the Barrow family in West Feliciana's plantation country, and an insight into the close ties binding many Southern families to their land.

POST-BELLUM FAMILY LIFE

THIS IS PRIMARILY A recital of my own personal experiences and impressions of my life. Memories I've kept alive, not all of them happy ones, but significant memories.

Mostly, I've returned to scenes that have given me delight, days of happy youth.

In each of our hearts, are tucked away somewhere in a secret place, memories that we have loved and cherished.

I have recorded these memories for those I love. Those who have known me as . . . .

"MEME"
HAZEL BARROW
August 17, 1953

I WAS BORN JULY 2, 1888, in West Feliciana Parish, Louisiana, on a cotton plantation.

There were five of us, Bird, Isabel, Ruffin, Eleanor and myself. Our parents' name - Bennett L. Barrow and Mary Isabel Barrow.

I was the youngest; I was given the name, Hazel. Mama liked this name because of a sweet little story she once read. I've always liked my name. I've learned that there was never any outstanding person with the name I bear excepting a famous "nut". I sometimes wonder about that, too???????

Our home was called "Woodland", an appropriate name for its location. It was on a high hill side, overlooking a beautiful sandy creek that ran through the pasture below murmuring a soft little ripple on its way into the creek. Large oak trees and some pecan trees made a shady and beautiful yard for us to play and romp in. Woodland was Papa's part of old Highland, the family home. Highland was built in 1799 by the first Barrow who came to Louisiana. We had no near neighbors except for our Grandparents and Aunts who lived at Highland. The place was quite isolated. Travel was by carriage, horse back and wagons. When the creeks would rise, one could not leave home as there were no bridges.

I was too young to remember much of our life at Woodland, but I have many happy memories of visits to Highland. When I was older and we had moved across the river to Pt. Coupee, Papa sold Woodland and we lived in St. Francisville a short period while a new home was being built across the river, in a plantation named "Preston".

Our Aunts and Grandparents, Grandpa John, Grandma Eleanor, and Aunt Mag, Nellie and Livy lived at Highland. We all felt that was our only real home. The Highland house was built with slave labor. It sat on a high hill too, overlooking a winding creek and green pastures. Its tall red chimneys could be seen many miles away. The yard was thick with moss draped oaks. All about the grounds were cape jasmine, crepe myrtles, sweet olive trees, magnolias and boxwood, a typical old southern garden. A wide brick walk approached the front of the house. There was a large front gallery entirely across the front. Four tall columns (square) across the front supported a high roof. There was no upstairs gallery. As you enter you come into a wide hall a lovely old hand carved door with transoms above and on each side small windowpanes. This gives a soft light into the room. Above the door upstairs is a palladium window overlooking the oak grove. The owls found this a favorite roosting place, and I recall many nights burrowing my ears in my pillow so I could not hear their hooting. As our bedroom was close by, they made cold chills run up and down my spine. I still feel a little queer when I hear those little creatures wooing and hooting at night.

The old cemetery is in sight of the Highland house. This is where most of my ancestors were laid to rest. Mama, Papa, and Bird are sleeping there. A quite peaceful spot, planted with sweet olive and jasmine, roses entwine the old brick wall that shelters and protects our loved ones from the outside world. In spring the birds nest in the old shrubs and trees where no one disturbs them.

I remember my first Christmas tree was in the big parlor at Highland. It was so beautiful. The morning was clear when we came running down the stairs and there was a gleaming tree lighted with real wax candles; gold and silver balls hung from its branches. It touched the ceiling and a ladder had to be brought in to reach the smaller gifts that were hung on the tree. I can never forget the joy and happiness of that morning. Santa Claus was such a real person to me. He had put a biscuit and a piece of meat on the tree for old Rex the dog. That really impressed me.

The Negroes on the place all came to the "Big House" (as they called it) to receive gifts of fruit, candies and maybe a piece of calico for a new dress, a toy for the little ones. They were very happy and contented people.

I loved Aunt Nancy, the old cook at Highland who often took me off to the quarters where she lived and fed me corn bread and "pot likker". She looked like a tiny old witch, her hands claw-like and wrinkled, but oh! What wonderful food, (pancakes especially) those old hands could prepare. Mama was sick a great deal after I was born and she spent many weeks in hospitals in New Orleans, so we were left at Highland during these periods. I was quite young when we moved to St. Francisville after our Woodland home was sold, I can recall nothing of being there. The old house we lived in is still there, owned by Eudolie Matthews and a most attractive place and comfortable.

Our home in Pt. Coupee was a very nice two -story one. Papa expected Mama to be well again and the new home would be a joy to us all. The house faced the river where we could see the steamboats pass. This was quite a novelty to us. The yard had a few pecan trees, but no flowers and shrubs like we had in the hills. We missed the lovely country roads deep with Cherokee roses that spilled their beauty and fragrance all along the way. In the spring, May and June, we picked the luscious blackberries, dewberries. Often this was our supper with rich cream poured over them. Blackberry cobbler was a favorite dish for the children.

We missed the dazzling colors of the West Feliciana woods in late October when the leaves were turning yellow and brown. We missed the dear Aunts and Cousins that we loved. Also the old Negroes at Highland who had petted and loved us. Many of these colored families followed us to Preston across the river. They loved "young Ma'ss Bennett" and wanted to be with Papa.

Our playmates were usually little Negros (we had no close neighbors). The quarters were close, not far from our house. Aunt Dicey-Ann, our old cook, was a faithful friend.

Mama came home to enjoy her new home (the home was completed while she was in the hospital) and there were so many things to be done. Gardens to be planted and there was great joy and enthusiasm in and about the new plantation. Ma, our other Grandmother, came and helped Mama who wasn't very well and unable to take over the duties of wife and mother. Mama was so happy to be with us again and so were we to have her again. I have a few old letters she wrote to members of the family. There seemed to be a note of sadness in them, and she must have felt a foreboding of the end of this happiness, for only a short time later the end came and dear Mother left us in 1891 at the age of only 30.

This was such a tragedy for us. Our new house was a saddened one. No Mother to care for five children. I was only three years old. Papa was a broken-hearted young husband. How different our lives would have been if she had lived. Papa managed as well as he could, I guess. Old Aunt Dicey-Ann was our second Mother. We went to her for advice and many times she settled our childhood spats. It was our great joy when we could follow her home to the Negro quarters and play with her children, Sissy, Lein, and a half dozen others. We did love her.

Papa decided we needed someone to look after us in the home, so he decided to employ a governess, Mrs. Reed. She was housekeeper, seamstress, and schoolteacher. I remember her being very kind and patient. I guess she had a difficult time filling so many places and dealing with adolescents and such as me. Bird was almost thirteen, so manly and dependable, loved by us all. He took over as much of the responsibility as he could. When Papa would be away, Sister, who had been known as a "tom-boy" felt a sense of responsibility and seriousness in our new way of life and developed into a rather serious person.

Ruffin, Eleanor and I did not realize what had happened to our happy home. Time passed on and eventually our lives adjusted to normal ones. Papa kept busy with the affairs on the plantation planting, keeping the Negroes peaceful and happy. I often remember a knock on the doorway into the night, then a voice "Ma'ss Bennett come quick, Tilda done hit me with an ax". Such things often occurred on Saturday nights when the Negroes would have their balls and whiskey, which was their popular drink. It would frighten us terribly, especially when Bird would have to go and settle their fights.

Eleanor and I were very happy depending upon each other for companionship. We played dolls and ladies or Miss as we used to call it. Most of our dolls were homemade by Mrs. Reed. There would be weddings and funerals, new babies born and real living among our doll families.

Often we buried a doll and dug it up the next day. Occasionally there was a baptizing in a mud hole under the old front gate, the kind we saw the Negroes have in the big pond, when they got religious. Such shouting and rolling and jumping. Our dolls would be almost ruined and it would be days before we could play with them again and a good scolding by Mrs. Reed. Aunt Eliza sent each of us beautiful dolls. I had never seen anything so lovely, and of course we were very careful with these dolls and loved them dearly.

I was always fond of cats. I remember so many I've had during my life, and as a child. I've had as many as twenty at a time. I played and dressed them like dolls and babies. I once dressed my favorite cat out in a new outfit, dress, bonnet and all. Tom did not appreciate or enjoy all this attention. He bounced out of my lap and made for the woods. Three days later he returned without his lovely new clothes. I was terribly upset over his absence and all the children were looking for him. He looked pretty guilty when he came walking home to get some needed food, but, oh! How happy I was when I held him in my arms again. When Papa sent us to the convent to school in New Orleans my favorite cat went along. He did not like it so he disappeared and that was the end of my cat playing. I really cried over the loss of my pet. This was one of my first sorrows.

Our lovely new home burned after Mama was dead about a year. This was another great blow to us. The house burned a few days after Christmas. Papa was away from home; the rest of us attended a Negro baptizing on the next plantation. We heard the old plantation bell ringing loudly and fast. We knew something was wrong because that was always a signal when something was wrong on the plantation. There were no telephones or other means of communication. Papa had returned and found the house full of smoke. Before much could be carried out, the house was gone, all our lovely old furniture, silver, jewelry, clothes and our beautiful dolls. The Irene family, on the next plantation and good friends of ours, gave us shelter, food and clothes. A broken-hearted Father and his children for once had to be separated.

Papa took us back to Highland, the old homestead, until he could find a place or build a new house. Adjustments had to be made for five motherless and homeless children. As young as I was, I can remember those heart-breaking days, leaving the home we had learned to love and separated from Papa. All our Aunts were young and it was not easy for them to have five children thrust upon them. Usually we were sent off to bed when the parties began. I guess we were homesick and missed our old friends across the river. We saw Papa seldom, as he had to look after the place. Aunt Livy was an invalid and could not take an active part in the social life at Highland. We spent lots of time in her room and she often told us stories and we loved her dearly. Eventually, Papa found an old farmhouse for us to live on the plantation adjoining ours so we went back to be with him. Oh! How happy we were to be with him and also to be in our own home again. Aunt Dicey was to cook for us and Mrs. Reed came back for a while.

Sister and Bird were growing up and I was nearing school age and when Mrs. Reed left us. We started to the public school in New Roads, a distance of six miles. We went in the buggy and the boys rode horseback. There were no school vans or hot lunches served to us and also no free schoolbooks. Aunt Dicey came at daybreak to cook our breakfast and prepare our school lunches of buttered biscuits, pork sausage, a sweet potato, and sometimes ginger-bread. Many times, the bad roads kept us or delayed us from school.

New Roads was a French community. We felt very strange and lonely. Eleanor and I would wander off where we could be together away from those hateful little Cajuns. I disliked their customs. Often, we traded our lunch with them. I loved bakery bread they brought to school. We had never eaten it, and our home cooked buttered biscuits tasted good to them. In winter we had a hard time keeping warm, traveling six miles twice a day to school. I still remember how numb I would feel. I wonder how we lived through these hardships and today how different.

Mr. and Mrs. Wills moved to the next plantation after the Irine family left. We were so happy to know they had a large family with children our ages, thus began a lifelong friendship between us and a source of great pleasure throughout our lives. On this plantation we spent many hours visiting each other. Mrs. Wills mothered our little brood who had no Mother and we all went to New Roads to school together. I can remember so well the delicious biscuits Mrs. Wills made. We often stopped by on our way from school to eat biscuits and buttermilk fresh from the churn. I took scarlet fever, and many times Mrs.Wills' biscuits were all that would tempt my appetite.

One of our greatest pleasures while living on the river was attending the Show Boats that came every winter. They were called "Floating Palaces". They landed at Bayou Sara and would play one night at each landing up and down the river. These shows were the only amusement country folks had, so they were the sources of many happy evenings. The boats played a calliope as they came near the landings. It could be heard for miles and such shrieks of joy when Papa would say, "Children we're going to the show tonight".

We had to cross the dark river in a skiff, how dangerous it seems now, but no sense of fear to us then, only the joy and pleasure we anticipated. I thought they were the most beautiful things in the world and it was the greatest joy we had ever had. The music, the colorful band in their gay red uniforms, the curtain rising and bright lights, singing, and dancing and oh! What beautiful girls and gorgeous clothes. I can never forget my childhood impressions of those old Show Boats, what a pity they are all gone.

Papa was especially fond of music. Our old piano had been saved from the fire. Many evenings we staged a show at home and our audience being the little negros, and sometimes the Wills children. The songs we heard on the showboats were the only ones we could produce. Birds' favorite song and a special feature was "Wish I Was Single Again". He often dressed up in Papas old "frock tail" wedding suit and danced a jig. This was always for an encore. Papa's favorite was "The Ship That Never Returned". The children would put a great deal of feeling and pathos in that very touching old Ballard. I didn't recall ever hearing it since I was a child and wonder sometimes who wrote it. We all thought it was beautiful then. I was too young to take an important part in these performances, so I would sit in Papa's lap and clap my hands.

Eleanor and I so often rode in the fields with Papa on his horse, one in front, the other behind the saddle holding tightly on. This was a great pleasure. Many times we were allowed to ride on the wagons loaded with cotton just picked and on its way to the gin house, there were large storage rooms filled with unginned cotton, where we would play and cover all but our heads with this fluffy cotton. We would come home with our heads, nose and ears filled with cotton lint. It was great fun on the plantation. So many simple pleasures. We lived so close to nature. I guess that is why all these memories of my childhood days stand out so plainly. They were so simple and such wonderful fun for us. We were happy and contented, never thought of a nickel or dime to run and buy cold drinks or gum as the children now have. Gum chewing was strictly forbidden and considered vulgar and common, especially in public.

The winter evenings at our house were memorable ones. I can see Old Aunt Dicey in the old kitchen cooking our supper, when we came home in the late winter evenings from school. The big wood fires sending out their yellow shadows and warmth into the lamp lighted room. The children gathered around the table in the dining room after supper and the schoolwork began.

Hog killing time was a great deal of interest to us. I can see the fires burning in the backyard, way into the night, where the negros would be melting down our year's supply of lard; a big skillet of freshly made sausages would send off the aroma of an appetizing meal, and the sausage would have to be sampled for proper seasoning. Hams and bacons were hung in the smoke house and for days the smoke would be pouring out. This was curing the meat and our dining table was supplied with this delicious food many months. It was a great privilege to watch all this interesting work done. We were called into go to bed and our noses were red as berries and hands stiff from the cold.

Our new home, built to replace the one that burned, was a modest one, not near as pretentious as the first. The kitchen was a separate building and the hall ran through the center of the entire house. A wide gallery across the front and back, four rooms opened into the wide hall. It was comfortable and we were happy.

Meeting the steamboats and watching the loading of the cotton was another pleasure we children always anticipated. The bales of cotton would be stacked out on the riverbank in front of our house where the boats would land and carry them to New Orleans for the world's markets. After it would be late at night when the boats came back, a big fire would be built for us to keep warm by and we played hide and seek on and around the cotton bales. That was the most fun of all. Many of a hard tumble I've had. Uncle Eugene worked on one of the boats that landed to take our cotton. We were always allowed to wait to see him if it wasn't too late. How majestic those boats looked as they came floating into the banks. The stage planks lowered a big search light that lighted up the landing where we waited eagerly to see it all. The Roustabouts, as they were called, began loading the cotton, singing and swaying as they worked. It was an unforgettable scene and one that stands out in my memory. Uncle Gene often brought us a bunch of bananas (this would be the only fruit we had ever had except at Christmas time). We did enjoy those bananas. I never thought I would ever be able to eat all the bananas I wanted.

A great sorrow came again into our happy little group. Bird, our dearly beloved one became ill. He developed pneumonia.

It was mostly fatal to have it after a short illness. He was taken from us May 11, 1896. Our dearest dependable Bird was no longer with us. Our protector, counselor, beloved by all, both white and black. Something in us all dies a little with the loss of a loved one. A quiet sadness settled upon us. It seemed unbelievable that Bird, only sixteen, was gone. We took Bird back to Highland where he was laid to rest beside Mama in the old family cemetery.

Our home was never the same after Bird left us. Papa was very unhappy. He could not leave us alone. As often his work took him away and there would be only negroes to look after us so he decided to sell our plantation and home and send us to the Dominican convent in New Orleans. This was a sad breaking up of our happy home. A terrible mistake and one that changed our whole future lives. I guess I was too young to realize what it meant. Sister and Ruffin perhaps felt it keenly. Never again would we all be at home together under the same roof, with no happy evenings together. Old Aunt Dicey caring and loving us, other old servants that we had learned to love, all this was behind us, a new world and future opening for us four homeless children. Even Papa was not to be with us now.

Our furniture would be stored away; even the dolls and little things were left behind. I recall quite plainly the early morning the boat "The Cleson" and Aunt Dicey coming to cook our last breakfast. There seemed very little excitement over all this. We could detect a tear in eyes. The old servants were all there to give us a hug and wave goodbye as the boat shoved out downstream for Baton Rouge, where we would then take the train for New Orleans. An entire new life before us, only memories now of our plantation home.

CONVENT DAYS 1897

AUNT ELIZA AND AUNT Sarah, Mama's sisters, lived in New Orleans. Ruffin was to enter public school and stay with Aunt Eliza while we three girls would enter the Convent. I was only eight years old. We visited with our aunts a few days before going to the Convent. I was thrilled with the big city. The big and beautiful stores, bright lights, street cars, these things we had never seen before. It was grand while it lasted. Our little hearts were quite happy again, but at last our visit ended and Papa took us to the convent. The gates locked behind us, Papa left and we were left alone among strangers. I had never seen a nun before. The dark black robes they wore, heads covered with those queer looking bonnets, beads dangling from their belts. Only a little white face peering out from that dark black costume. We went into this quiet atmosphere of gloom that prevails in a Convent. The dull funeral air frightened us. A bell rang solemnly in its distance (one does everything in a Convent by the toll of a bell). How differently it sounded from our gay old plantation bell that called the workers to their rest hour at noon and from their daily toil at eventide. I wondered if I would ever be able to laugh and romp again.

Oh! Why did Papa send us to this place? What kind of a house was this? I cried, Sister cried, Eleanor cried. It was hard for me, a child of eight, to realize our home was now a Convent and when summer came where would we go? I did not know what plans Papa had when that time would come. I guess this was the most unhappy time of our young lives and one that still stands out in my memory. The older girls had separate rooms from the small children. I was allowed to sleep beside Sister, a request

that Papa had made before going to the Convent. At daybreak
every morning, a nun awakened us by walking up and down
the corridor ringing a bell. We often saw the stars still shining
on our walk to church which was every morning. Oh! How I
did hate to get up and I can never forget how cold I would be,
no fires anywhere until we reached the classrooms for the days'
work. I practically lived with a cold and grew thin, and pale. We
seldom saw any sunshine. I think the nuns were a little worried
about my health and tried to give me some extra nourishments.

Aunt Eliza and Ruffin came to visit us sometimes on Saturdays
or Sundays. These were the bright spots during those days.
Aunt Eliza was so good and kind, looking after our needs, buy-
ing clothes and keeping in touch with us as much as she was
allowed. We were permitted an occasional visit to see her. The
remembrance of our first Mardi Gras is unforgettable. It was
the loveliest thing we had ever seen. A world of make believe.
The beautiful floats, bands with their lively music, the mask-
ers throwing trinkets from the floats as they passed the crowds.
The night parades were the most beautiful, torches flaming to
light the way. It was a spectacle I had never seen before.

We seldom saw Papa. He had employment in a distant part of
the state. We missed him, but gradually became adjusted to the
routine life at the Convent. There were many nice girls that I
became very fond of. I was in music and many other school
activities. I still never liked the silence and could not see the
necessity of it. It was too solemn for an eight-year-old. I truly
felt I was an orphan.When June came and school was dismissed
all the girls were gay and happy over going home. Then we
realized we had no home to go to. It was somewhat of a shock
because we hadn't anticipated what it would mean staying in
the Convent all summer. How could we endure it? A feeling
of sadness came over us as we saw the girls leave with happy
young hearts. Then it happened: We got an invitation from rel-

atives to come and visit for the summer. We were going back to West Feliciana. Oh! How happy we were. We were going to travel by boat.

Aunt Eliza and Uncle Gene took us to the boat "The Natchez". It was a wonderful trip. The captain knew Papa and he allowed us many privileges. One was a trip up to the pilot house and a hand on the big wheel that steered the boat. We rang the big bell that signaled the engineers down in the engine room. It thrilled us to death.

Our summer was very happy. It was grand to sit down to the table, laugh and giggle as children usually do. Play in the woods, make the creeks and visit our different cousins, gather the eggs in the evening and watch the cows being milked. It was heavenly. During the summer the yellow fever broke out in New Orleans. Quarantine was set up all through the state, travel was at a standstill. That meant we could not return to New Orleans. Cousin Calhoun Hamilton had us stay with his family and we attended a small public school at Laurel Hill, a little red schoolhouse sitting back in a grove of magnolia and beech trees. Only one teacher, Miss Jane Augue, taught all grades. I don't remember how many children she taught but there were about ten of us from one house. Cousin Calhoun and Cousin Dora had ten children, their grandmother lived with them, and we were four and so altogether seventeen.

Their home was not a large one, but there always seemed room for one more. Everything on their table was homegrown, meat, vegetables and all. I can never forget those "beaten biscuits" we had for supper. They were thin and crisp, a special favorite for Cousin Calhoun. There was an apple orchard and home-grown peaches and watermelons. Dog in the Blanket was an apple dish. We all loved Cousin Dora. She was a wonderful wife and mother ruling over that house full of children like a queen.

Everyone loved her and respected her word. She made every piece of clothing worn by her family including the boys' shirts and some of their blue jeans. It was the first time I ever saw little homemade moccasins for babies made from the soft tops of high-top shoes. They looked like a shoemaker's job.

When the late evening came in and the lamps were lighted, all of us came and gathered around the big open fireplace often Cousin Dora would tell us stories, and many times the old flat piano would be opened and we would gather around and sing while she played. Sometimes we danced and had as much fun as though it was a real party.

I loved Cousin Dora very much and I guess these were the happiest times we had ever spent since leaving our own home across the river. We rode to school in a small wagon, a distance of a little more than a mile. Cousin Dora prepared a large basket filled with goodies for our lunch. A gallon of milk brought along in a brown jug. We put the milk in a spring down behind the school house. The spring water was cold as ice and trickled down the hill side and this kept the milk cool and fresh for our noon lunch. My, how good it did taste with that lunch. All the children sat under a large magnolia tree and ate together. It was like a picnic every day; even today I imagine you can find the names carved into that old tree of boys and girls who went to Laurel Hill School. That was quite a fad to cut your best girls or beaux's name beside your own. Another summer passed and then it was time for to return to the Convent. Ruffin did not return to New Orleans, but went to work with Papa, who was building levies for the state. In those days, they used convict labor. The levies were built with mules, no machines to haul dirt then. It was a slow process and hard work. We missed Ruffin and saw little of him and Papa for several years.

It was not hard to adjust ourselves to Convent life as it was before. We had learned to take the hard knocks and face life with a better understanding. Cousin Dora had taught us many things that we had missed not having our own mother. Several years passed and we remained in the Convent, our summers were spent in the Convent. It was hard seeing the others leave, and we had no home to go to. Aunt Eliza often took us to visit with her. We enjoyed playing with children in her neighborhood. I was amused hearing the old buttermilk man calling "buttermilk" every morning. The old rag man called out "Rags-rags, anybody got rags?" We played out on the avenue under the big arch light and always there was the old "Praline Women" selling pecan and coconut pralines and pink and white taffy candy. She got most of our nickels (There were very few of them, I can assure you).

When we returned to the Convent after these visits, I remember being very lonely and wishing I had a home like the others. No matter how kind the Nuns were to us, it wasn't like being loved and having a mother. Sister finally was finishing school, time to graduate. Eleanor and I were left to battle alone. A cousin, Mrs. Tom Glynn, who lived out in the country from Covington, La. offered to give Sister a home, she needed someone to be with her and several small children during her husband's absence. Sister was to be a helper with a very small salary. She had some very interesting experiences over at Cousin Mary's home. It was far out in the country and not near neighbors. The climate and pure country air was what Sister needed after being shut up in a Convent. She had grown so thin and the duties of caring for chickens and gardens and other chores on a small farm seemed just what she needed, Though it was often hard, she benefited by it; there was plenty of wholesome food.

Cousin Mary invited Eleanor and I to spend one Christmas over at Covington while Sister was there. We were thrilled to death to be going somewhere like the others were. The train ride delighted us. We were met at the station, then a long ride into the quiet countryside by buggy. The tall pine trees cast shadows along our way through the deep woods. I thought it was beautiful. It was night by the time we arrived at Cousin Mary's home and very cold. How wonderful it was to again stand in front of a big wood fire and warm ourselves, and sit around the fire and laugh and talk.

Our visit was delightful. All of us went into the woods to get a Christmas tree, and Oh! How wonderful it was to breathe the fresh cold air. The murmur of those pine trees remained with me a long time and Christmas morning there were gifts for all of us. Simple little trinkets, but what happiness it brought to our little hearts. I still remember a little pocketbook with a handkerchief folded in it and a great big silver half dollar all for me. I don't think I could have been any prouder if it had been a whole dollar or finer purse. Our visit ended and we were driven to Covington to take the train. It was a late winter's evening and Eleanor and I got off the wrong place. When we were nearing the city- it was on the outskirts of New Orleans- we realized our mistake when we saw the train going down the track out of sight.

I don't think I can ever forget that feeling of loneliness and fear. No one in sight, darkness all around us, we took our bundles and bags and began walking. We decided the best policy was to keep on the railroad track. There were no houses near. I guess God was watching over us that awful night. We cried and prayed. Two terribly frightened little girls. After walking quite a distance we saw lights and heard the rumble of an electric streetcar, so we decided to take the first car we saw as we knew we would be safe on a car. We made three transfers before we arrived at Canal Street where we could take another one to Aunt Eliza's house. Dear Eliza was so precious to us and her son,

Gene, and his wife, had gone to meet us. Her eyes fairly popped when she heard our story. We all had a good laugh when Gene arrived, though it wasn't a bit funny till it was all over and we were safe for the night. I don't think Eleanor or I can ever forget our visit to Covington.

Papa finally got a job where he could send for us and we could spend the summer. He moved to Angola and no longer did levee work. Angola is a large penal farm and cotton was the principal crop. We boarded with Mrs.West Reilly and her family. We went to Angola by boat; it was named "The Valley Queen". This was another happy period in our lives. The boat was filled with boys and girls going home on vacation. We played games, had delicious meals. The state rooms had bunks, one on top of the other. I had never slept in these before and I enjoyed every minute of this trip. Another blunder occurred. No one was there to meet us.

It was about midnight, a lonely boat landing, only a convict watchman on duty, we had no idea he was a convict. He occupied a small one room cabin where he slept. He was very kind to us. Of course he knew Papa well and said he was not expecting us until the next trip of the boat. Anyway, he let us lay on his bed to rest during the remainder of the night. The mosquitoes nearly ate us up. I thought daylight would never come. Again God was watching over us, how could we ever doubt it? The sun finally came up and, in the distance, we could see the convicts going out to work, their hoes over their shoulders, singing songs as they went into the fields. It imprinted in my memory this unusual spectacle. It was a fresh cool June morning, the grass heavy with dew, morning glories blooming over the countryside. Here everything was real things of beauty that God had made. No paved streets, Convent walls, clanging of bells, things we had lived with so long. Here was a place for us to rest and dream the whole summer long.

Mrs. Riley was a dear good friend, they had a family of boys and girls and we saw Papa every day, so our experiences were all happy ones. There were horses to ride, we fished, and danced and I guess this was one of the happiest periods of our young lives. Papa had a convict band and often brought them to play for our dances. Sister left Covington and joined us at Angola, and Ruffin was now living St. Francisville working in the drug store. He came to see us several times and we were altogether for the first time in many years. Papa left Angola at the end of the summer. The Riley's were leaving too and moving to Tunica, so we begged not to be sent back to the Convent. Mrs. Riley and family let us move to Tunica and we shared a small home where their old parents lived. A dear old couple. We could not live alone, so this was a very desirable plan.

I entered the public school at Tunica, Tunica Hills as it was called, was the most picturesque place I had ever seen. Our little house overlooked a green valley sitting high on the edge of a hillside. The house was a dog trot house, an open hallway through the center, rooms on each side of the hall and a gallery all across the front. The kitchen was in a separate house, and also the dining room. All of this was up an elevation of about three or four feet from our sleeping and living quarters. It was not so nice in winter, but very delightful in summer. Our old cook, Aunt Julianna, cooked many delicious meals on a little wood burning stove. She lived close by in a little cabin. The chimney was made of clay mixed with moss that held it together. The walls of this house were heavy hand-hewn logs, clay between them to keep out the cold air. The floor was a dirt floor and I remember so well how we loved to go there and sit before that old fireplace and talk with Aunt Julianna. The smell of that old clay pipe she smoked did not bother us one bit. There would be only the light of the fire to see by, but such a cozy scene and we delighted in sitting on the floor and listening to the many tales she would tell us.

All these simple things are imprinted in my memory and represent happy days of my childhood in Tunica. I was happy as a child should be, the people were kind and friendly and as my memory goes back to those days in the shadowy past, those people who crossed my path stand out on the screen of memory. So often, I wander back in my thoughts to those days. My school days in Tunica also recalls many happy times. I had to walk two miles to school through a wooded section. No girl would be allowed to do that today. I was not afraid nor was it considered dangerous. We had school in a small church building. We had only one teacher. It was a lovely spot right in the woods. Grape vines grew in the trees and we used these for swings. A favorite past time at recess was to see who could swing the highest out over a deep ravine. We never realized how dangerous this was. Those vines were thick and heavy but could have broken anytime.

I had some very dear friends in this little school. Many weekends we had horseback riding groups and parties and hayrides. I had one special friend Ada Coates who often walked with me to school. Ada's home was a favorite stopping place on my way to and from school. She was an adopted daughter of an elderly couple, Mr. and Mrs. Porter. It would be hard for me to describe the beauty and quaintness of this spot where they lived. This picturesque setting was or could be an artist's dream. An old well house stands out in my book of memories. It was here I had many a cool drink of water on a hot summer day. The old wooden bucket was made of cedar and a gourd dipper to drink from.

The roof of this house was wood shingles so old that green moss was growing on them. A large front porch shaded with oaks.

The rooms were large with a wide hall running through the center. Mrs. Porter had a little organ that she played and I still recall hearing her playing and singing as I came over the hill. One I remember so well was "Long, Long Ago". I will always associate that song with this lovely old lady. Both Mr. and Mrs. Porter and also Ada are gone from this world. The old house was torn down many years ago. Miss Ethel Jones, who was the older sister of my best friends, Holley Jones, often took me to school in their surrey. It was a longer route but saved me a walk, and after a time I discontinued my walks through the woods and road altogether. Miss Ethel picked up all the children along this route, so sometimes there would be eight of us in the surrey packed in like sardines. We had lots of fun singing and talking on our way to school. She came back in the afternoon and brought us home. I missed the walk through the woods, especially in the springtime when I often stopped and watched the birds flitting about. We had lots of fun singing and talking on our way to school. I've always loved birds; I still do, and try to attract them around my home by feeding them and keeping a bird bath around.

After spending several years in Tunica, Papa decided we would have more advantages in St. Francisville, so we said goodbye regretfully to our dear good friends, I felt a keen sense of sorrow in leaving this place where I had been very happy, the only home I had known for so long. I had learned to love the woods and hills, the call of the Whip-o wills in the late evenings, the singing of the locust. I will always associate those sounds with old Tunica. The nights there had a million voices, it was as if I had an ear for every sound. The moonlight nights flooded me with a warm feeling of kinship for all the things around. I felt myself a part of all this. These invisible choirs of night, I will always remember. My heart ached the night we went aboard the "Betsy Ann", a mail boat that ran between Bayou Sara and

Natchez. Tears ran down my cheeks as I looked back feeling myself drifting out into the dark muddy Mississippi entering into a new period of my life. I was then fourteen years old.

We lived with an old cousin, Cousin Margaret, when we first went to St. Francisville. Later we found a house where we could all four be together for the first time in many years. The house was an old one, yard nice and had large oak trees around it. We had some of our old furniture from across the river, including the old piano. We were a very happy family again. It was wonderful to have Ruffin with us again and what a joyous time we all had fixing up our new home. We were all quite happy again, the past was behind us. A new future of anticipation lay before us. We had our old friends of childhood days and fond Aunts and Cousins; our social life was all we could desire. Again we went to Laurel Hill to visit many weekends going up on the old Y&MV train, more popularly known as Mr. Argues train. He was the only conductor for many years and loved by all people along this route.

Cousin Dora would send for us in the little school wagon. Often it would be a cold winter evening or perhaps night and a big bonfire would be burning as we got off the train where they were waiting for us. It was a happy reunion for us to be back at Tanglewild. Cousin Dora played the piano for us to dance. Uncle Alex would bring his fiddle and they would be our orchestra. Those were truly happy times.

Ruffin and Eliza, (Cousin Dora's oldest daughter) had been sweethearts for many years, so they became engaged and made plans to marry and come to live with us. On December 16, 1904, Ruffin and Eliza were married at St. John's Episcopal Church at Laurel Hill. I thought Eliza was beautiful, and Oh! That dress of lace and satin. I have never seen many brides, but I couldn't imagine anything lovelier than Eliza was. A big country dinner was arranged for all who came to the wedding. It was at noon. I well remember how ill I was from that big feast. We had an interesting time fixing up the place for Eliza. Eleanor and I gave them our room and moved upstairs.

A month later Sister announced her engagement to Tibbie (Charles Thibodaux) whom she met at Angola while we lived there, (our home was gradually changing) so she and Tibbie were married in the Catholic Church in St. Francisville in January 1905. Everyone helped to make it a very pretty wedding. We were carried away with all these exciting events. Papa came home to give Sister away and visited us a while. Sister went to Angola to live as Tibbie had a nice little house ready for her. Eleanor and I continued to live with Ruffin and Eliza. I was sent to Woodville, Miss. to boarding school. It was known as Edward McGhee College and several of the girls I knew were students there. It was a small school for girls. I loved the college life so unlike those convent days. The people of Woodville were so nice to us and often entertained us at suppers or parties. I cherish the memories of those days spent in Woodville, the close friendships that I made there, some whom I shall hold dear.

In another year Eleanor married her childhood sweetheart, Joe Wills, the family we had known across the river. We had never lost contact with our old friends through the many years since we left Preston and spent many days visiting with them. Eleanor and Joe were married in Grace Church, January 19—. So she too left and lived again in Pt. Coupee, next to our old friends and plantation home although they did not stay there very long.

The next big event in our house was a baby. The first one too. Montrose was Ruffin and Eliza's first. I'll never forget how very cold it was. A coating of ice covered the trees. Everything looked like a fairyland. so unusual for our climate. We had open fires to keep warm, and drying baby clothes was quite a problem. We thought Montrose was wonderful and made quite a pet of him. Two other sons, Alston and Ruffin C., were born to Ruffin and Eliza

Papa left about this time to work in Panama where he remained a year or more. Then he returned home and married Edna Floyd of Delta, LA. They had met while Papa was on a construction job there and I guess that is what drew Papa back to the States. We were very happy over Papa's decision to marry again. He told us of Edna and though we had never met, there were many letters exchanged and we felt that Edna would make him happy. Papa had wandered around so long and needed a home and someone to love him. They were married in 19—and came to visit us and meet the large Barrow family. We all liked Edna and a new era of life began for him.

I was the only one left unmarried, so I made my home with different members of my family. Most of it was with Sister who lived in Bayou Sara; she was made postmaster there while her husband worked with Irvine Grocery. I was appointed Asst. postmaster, my first job and a much needed one as Papa now had another person to care for. I was very happy there and cherish the memories of my days there that were so carefree. I had youth and health, but so little of worldly goods, the little home was humble but what did that matter? There were beaux's parties and visits to friends and all the world was rosy to me. In this little country post office I saw for the first time the man I was to marry.

One hot summer afternoon no one was stirring; the streets were deserted. I was sitting there dosing and I guess quite bored. A stranger was quite a novelty in those days. Two nice young men walked up to the Post Office window and asked for their mail.

''What name please???'' "Gordon and John Barrow." Well, my name being the same, my eyes bulged and my heart beat a little faster. I was quite confused but tried to keep calm and not appear stupid. I thought it was a joke and some smart alecs trying to get acquainted. I knew all the Barrow families, far and near. My thoughts were in a whirl. Where did these Barrows come from and why were they here? And asking for mail? In those days girls did not talk to strangers, so I was cautious as to what I said. Well, I soon found out about the two strangers. Both were pharmacists and had come to St. Francisville to work in the two drug stores in town. Well, we've often laughed about that experience, as I married one of those strangers a year later.

Our courtship was a happy time, a bright milestone back along the road of my life. I remember myself thinking of him and listening for his footsteps each evening, counting the hours we could be together. The evenings he did not come, the mail would bring a tiny box of violets, my favorite flower, a note or some little token to let me know he was thinking of me. Many evenings we sat out on the levee watching the boats pass or he would bring his rifle and shoot turtles and snakes along the banks of the river. It was during these days, we planned our life together, our dreams and ambitions. I treasure the memory of these happy times, the many thoughtful things he said and did. All through our married life, those same thoughtful characteristics remained.

As he was only employed temporarily in St. Francisville, the time came too soon when he had to leave and our happy times being together each evening had to end. When he left me, there was a diamond ring on my finger. Papa had been asked as was customary those days, and our engagement was announced. Gordon went to Memphis where he lived with his Mother and Sister. He was employed by a large drug firm, Jas. Robinson, but after a few months of city life, he decided to go to a small town (Wilson, Arkansas). I think the lakes and prospects of hunting and fishing attracted him to Wilson. So that is where I began my married life.

After Gordon left St. Francisville, I continued to live with Sister and after several floods there, Sister went to join her husband who was doing construction work on levies. Ruffin took over the Post Office work and I went to live with Papa and Edna in Baton Rouge. Papa had charge of the receiving station, better known as the wall for the receiving of prisoners. They had a very comfortable place; part of the house formed the wall of the prison and it was cold and very large and roomy. The hall upstairs overlooked a courtyard where the convicts sat or walked during recreation hours. It was also the place they entered when being sent from the outside world.

I have often seen fine young men enter those gates hand cuffed, well-dressed, looking sad and dejected. A short time later, they would be seen with their new outfits, stripes, head shaven, a large black number across their backs, their freedom gone, and a new life before them. I've seen many sad scenes from that window upstairs when wives, mothers and children came to visit their loved ones. It depressed Papa quite a lot as he was very tender-hearted, though he was fearless and a stern man. The men were fond of Papa as he was kind and always had time to listen to their pleas. An old Chinaman (called Wong) was a faithful dining room servant for our family. His devotion to Papa was touching and especially for Floyd, our little brother, who was born at the walls. The first summer I was in Baton Rouge, Gordon came to visit us and we made definite plans to marry at Christmas, December 28, 1909.

I had very little to buy a trousseau with. I had saved a few dollars from my Post Office job, and some help from members of my family and Papa. Fortunately for me life was very simple in those days. People wore homemade clothes, but in the end, I was fitted out very nicely and had as much as many other girls had. I was married in a lovely, tailored suite of brown broadcloth trimmed with heavy moire, silk collar and cuffs, flesh colored georgette crepe blouse. I wore high-topped button shoes. The buttons were pearl, the uppers were cloth. My hat was beautiful, a large brown velvet with a flowering brown and orange lip plum falling to the side and around the entire crown. I had never had a hat so lovely and outfit so lovely and felt like a queen. (Poor little girl, so very, very happy) I wore a cheap fur neck piece and carried a muff. I had never had riches, so my outfit was all I desired. I only wanted the one I LOVED. I knew I would be well provided for and Gordon loved me devotedly.

We were married at 9:00 A.M. at the Grace Episcopal Church. The Church was filled with interested friends and loved ones. Christmas greens and a few flowers made the old Grace Church look lovely. The early morning sun shone through the old stained-glass windows throwing many colors over the soft green carpet and over the candle lit alter. The golden mellow notes of the organ sounded around through the old church and Papa as I entered. Gordon stood at the foot of the chancel with John, his brother, and our Episcopal Minister, the Rev. R. R. Claiborne. Some lady told me later I was the happiest looking bride she had ever seen. My face radiated happiness. I think this was the happiest clay of my life. I felt that I belonged to someone. So much of my heart in youth had been spent in other people's homes amid other people's possessions. My deepest longing had been for a place of my own. However humble, my hopes and ambitions were to make it reality. I admired and loved the tall handsome stranger who came into my life one summer afternoon in a country post office.

We left after the wedding ceremony and went by train to Memphis, Tennessee to visit Gordon's Mother and Sister who gave me a hearty welcome and made me feel I belonged in the family. I grew to love them very much. Gordon's young Uncle Robert Gordon and his bride came to visit at the same time. They had been married the day after we were. Bob and Gordon were like brothers and raised together so it was especially interesting for us all to meet and be together. Those few days in Memphis were wonderful. I enjoyed meeting Gordon's friends, going to theaters, and I treasure the memory of those happy days.

I wondered if I could ever like Wilson, Arkansas, it seemed so desolated and barren, I missed the old oaks with its grey moss. It seemed away from the civilized world. We boarded with Mrs. Oliver, such a fine lovely little person, and so kind and good to me, as I was often lonely and homesick. Mrs. Oliver had other boarders and also a large family to care for. I had never seen anyone who could accomplish so much with no help. There were no negro servants like I had been accustomed to, and I wondered how she did it. Her little boy died soon after I came there to live, and it saddened me so very much. I had often cared for him while Mrs. Oliver did her many chores. A sweet little fellow and a very pretty child.

Gordon worked long hours at the drug store. I was alone so much, often late into the night. How I envied the other girls whose husbands came home at six o I clock. I was pretty lonesome while he was away. I missed my family and friends back home, but I soon made new friends and came to like Wilson very much. I became interested in church work and other small-town activities and found myself adjusting to my new life. I grew so weary of boarding, so a house was built for us, our first home. I recall what a happy time we had selecting pieces for our home, and our many trips to Memphis on shopping expeditions. Our little home was a palace in my eyes. How proud I was of it. I was not a very good cook so poor Gordon had to suffer from that. We had many laughs over some of those biscuits I cooked. Mrs. Oliver was usually my counselor and helped me with many of my housekeeping problems (she lived next door). We spent many weekends in Memphis. Gordon wanted me to see all the good shows. Pleasures I had never had before.

An epidemic of smallpox broke out in Wilson not long after we came there, and I was then with Mrs. Oliver. Gordon sent me to Memphis, and while I was away one of the members of Mrs. Oliver's family took smallpox right in the house we were living in. It frightened us terribly, as we all had been exposed. I remained away several weeks and Gordon moved to another boarding house. When I returned, I thought I would never find my belongings.

Wilson was a low swampy place, so Gordon and I developed malaria. I felt wretched most of the time and he had to remain away from the store quite a lot, so we decided we had better leave and look for a place where Gordon could have his own store. I rather hated to leave my lovely little house. I returned home for a visit; it was wonderful to be back in my old home-town and visit all the folks. Ruffin, Eleanor and Papa were in Baton Rouge. Sister still living back in Bayou Sara and took over the Post Office again.

Gordon bought a store in Wellington, Tenn. only seventy miles from Memphis. Wellington was a pretty little town. I liked it at once. We boarded with Dr. and Mrs. Chambers. They were a nice happy family. We stayed with them until we could find a house. We were not so fortunate this time. We had to take what we could get. The house was much too small for our big heavy pieces. We had no conveniences. My things looked so out of place in the tiny rooms, but we were happy in spite of the obstacles. I had to cook on a small iron stove that burned wood. I spent half of my time nursing burnt fingers and wondering where I would put all my things. We wanted so much to build a home but knew that we would have to wait until the business was well established. Sister and Eleanor visited us and John, Mother and Sis came often. I made some very good friends in Wellington, after we were able to have a car, we enjoyed trips to Memphis. I took part in the social life.

Gordon and I often were chaperones to groups of young people. I enjoyed the church revivals there. We had never had revivals in my hometown. I found them very interesting and often quite amusing. A large tent with sawdust on the floor and wooden benches would be used as the church setting. It could not hold all the people who came, people in wagons, buggies and horseback. They brought their dinner and babies along. This was all very bewildering to me and I would not have missed it for anything. Often people became offended when the preacher became very personal. In one town nearby there was a shooting affair at one of these revivals. There always seemed to be an air of expectation at these meetings. One never knew what the preacher would say.

A great sorrow came to us while in Wellington. Gordon's sister, Virginia, (whom we all called "Sis") died very suddenly. She was ill only a few days. We felt stunned over this great tragedy. Sis was only twenty years old; she possessed great qualities of charm, vitality and true passion of gaiety and life, a most attractive person, and remarkable talent for music. She and I had become great pals, and a terrible vacancy was left in Mothers little home in Memphis. All the sunshine seemed to have left at her passing. Mother took this sorrow quite bravely at first, but she began to break under this terrible strain. Covering up her sorrow soon began to tell on her. The loneliness and despair could not be hidden. John came to live with Mother where he was employed as a pharmacist, but his health began to fail, confinement hastened on an old ailment, so he had to leave, going to Florida to seek better health. Mother was left alone. She prepared to remain in Memphis among friends and did not break up her home as so often people do when a death occurs.

Gordon and I made plans to build our home. His business was good and we grew tired of moving about. Our home was a bungalow type (so popular at that time). I didn't think any couple could have been happier than we were when the time came to move into our new home. It was all we desired. We had enjoyed daily watching its progress and completion. We moved in June 1914 and on July 5, 1915, our first baby arrived. Virginia was born in the new house. These two great events stand out in memory. Virginia was named for "Sis". Mother begged us to give her that name. A name very dear to us all. Finally "Gin" was the nickname that Gordon gave her and one that has stuck through the years.

"Gin" was a beautiful baby, a head of brown curls and so very like her Daddy, which pleased him immensely. She became a neighborhood pet and I was never at a loss to find someone to keep her. A happy contented little girl. Mother thought she was the only baby in the world and visited us a great deal. I was kept very busy with my new home and a baby to care for. I enjoyed every minute of those interesting days. Our happiness was complete. The joy and pleasure of the newborn baby and home for both of us will be forever imprinted in my book of memories.

Again sadness entered into our lives, Mother met a tragic death when Gin was only seven months old. Mother was badly burned by lighting a gas heater. She only lingered a few hours and then she too was gone. It saddened our lives beyond words, our visits to Mother were at an end, and the little house was sold. I packed Mothers personal things and things we were to keep. It was a heart-breaking job. Our hearts were crushed and saddened and only memories lingered of a happier day when Mother and Sis were with us. A happy honeymoon visit and other good times we had had were all behind us.

Gordon and I seem to wonder listlessly what to do, there seemed so little to keep us there in Millington, so we began to think and plan to go back to Louisiana. We loved Louisiana best and had hoped someday to go back home. Gordon bought a drugstore in Biloxi Mississippi and we moved there in 1916. Gordon loved the fishing and water, and we would be near home. Again it was bidding goodbye to the dear friends I had grown to love; our sweet little home was for sale. I've only been back once since I left Millington. It's now quite a city, and I often have a longing to go back and take a glimpse of our little home.

We had an attractive cottage on the beach in Biloxi. I loved to watch the tide come and go and hear the call of the seagulls in the early morning as they dipped into the water hunting for fish. The yacht club was out in front of our house. Over the water it was interesting to see the lovely boats come in and dock there. Sometimes the waterfront would be thick with these beautiful yachts owned by wealthy vacationers. Gordon was too busy getting his business started to enjoy the fishing and swimming he had planned. Robert Gordon and Kate, the uncle and aunt I first met on our honeymoon in Memphis, came to visit us in Biloxi.

The worst storm I had ever been in occurred while they were with us. We were marooned for days while this hurricane swept over Biloxi. Everyone kept indoors, no groceries were delivered, we were almost out of food, and two babies needed milk. Bob and Kate had a baby girl, Elle. What a time we had keeping those little tots happy. When the storm subsided, there was debris everywhere. Beautiful yachts with only their tops showing lay out in the harbor, some turned completely over, the yacht club was blown away, the wind had blown water under my doors before I could take the rugs up. Therefore they stayed wet for days, soggy and mildewed, the furniture warped and ruined from dampness. Mosquitoes came into Biloxi by the millions. All the visitors began to leave. Biloxi became almost a ghost town, our business was going on the rocks, the strain was terrible on Gordon seeing his prospects and dreams vanish. He sold the business at a great sacrifice --- so again we were on the move.

We went to St. Francisville and Bayou Sara for a visit where Gordon enjoyed a needed rest. We sold our furniture and stored our personal things until our plans were made to locate somewhere. Uncle Sam, Mama's brother and one of my favorite uncles wrote Gordon of an opening in Beaumont, Texas. He was so anxious for us to locate there. So we went to Beaumont, as soon as we could find a suitable place to live. Gordon worked in a suburban store, a very nice place, our little apartment was small and cozy, on a busy street corner of a nice residential section.

I had never lived in such small quarters before but being wintertime I liked it. Gordon bought a bicycle to go to work on. I remember how comical he looked riding off in the early mornings. Gin and I would wave to him from an upstairs window (as we lived upstairs). I had so little to do, so my whole day was given to Gin. We took long walks, morning and afternoon. Often going window shopping and to picture shows. I took her to see "Snow White", then playing, and played by Marguerite Clark. I had told her the story so many times so she got quite a thrill over this picture. We saw many of Mary Pickford's pictures.

She was then such a cute little girl at the height of her career. We three were very happy in Beaumont and though I did not meet many people, I had many interests to make me happy. I so well recollect the cold winter evenings, the Hot Tamale man blowing his funny little horn. It sounded somewhat like a flute. Gordon often came in with a package of them for our supper. I realized now that may have been the beginning of a very serious gall bladder ailment for, I finally ended up having an operation. Our first Christmas tree for Gin was in our tiny apartment in Beaumont. It was an extremely happy Christmas, just the three of us, most of the toys came from Kress -- but a small child knows nothing of the value of a gift so Gin was over-whelmed with her tree. Gordon and Gin had talked to Santa Claus for weeks before Christmas. He was determined that the old tradition of Santa would be a real thing in Gin's life; letters were mailed to Santa, a cup of coffee left on the stove for Santa. All these old-time customs amused her very much; it's hard to remember who had the most fun, Gordon, Gin or I.

Uncle Feltus (Papa's Brother) wrote Gordon suggesting he come to St. Francisville and open a drugstore; there was only one in town. Well, that letter made a great change in our lives. We wanted so much to return to St. Francisville, so here was our chance. Our hearts were over-flowing with joy and enthu-siasm over this opportunity. A new future lay before us going back home - the dream of joy of which we had waited so long. Maybe this would be our last move (It was). I had no regrets living in Beaumont, but we had never considered it a perma-nent home.

We lived with Sister and Tibbie a few weeks in Bayou Sara while Gordon was getting his store repaired, goods assem-bled and the business started. Sister was again running the Post Office, Tibbie operated a ferry on the Mississippi River, all the children in our family loved to visit Sister and ride on "Steamboat Bill", so of course Gin too was thrilled to ride and watch the many amusing things that happened on the river. We soon found a small house we could rent; it was near the store. I had to buy new furniture as we had sold most of our pieces when we left Biloxi. The house needed repairs badly and had few conveniences, so Gordon and I fixed it up very nicely ourselves. I painted and worked for weeks making it attractive. I had a little negro girl to keep Gin and take her out in the evenings and meet the other children around town. This was really the first playmates she had ever had. It was grand to be back home where we knew everyone and renewed old friendships of younger days.

Papa, Ruffin, and Eleanor lived in Baton Rouge. Sister and I were only a mile apart. Our old friends seemed happy to have us back. Gordon, who had only lived there a short time before we were married, had made many friends and his business began to prosper in a short time.

After a year in St. Francisville, we felt this would be our permanent home. There was a picturesque old house of brick only two doors from the drugstore which appealed to us tremendously when we were looking for a home to buy. The gentle charm of this old place, with its large trees, wooden gates and fences, enhanced by old cameleer, rose bushes growing in the yard, had the lived-in quality; the sense of love lived within its steady old walls, fire places that had warmed another generation, a house of homespun living, perhaps, hand hewn slippers, copper toed boots, or tiny satin slippers had worn its toes in front of those old fireplaces, a dog, a cat had curled up in its hearths; thus we bought this old house.

In June, we moved in after months of repair work was completed. I loved every brick and plank of this house. The rooms of plaster gave it very effective walls a foot thick, long wide porches that we both loved. The house sits on a high hill, overlooking the river a mile away -- we could see the boats as they came around the bend of the river. It was truly our idea of a home. Nothing finer could have appealed to us more. We had a large, screened porch upstairs, this is where we slept in the summer, and an elm tree shaded this porch from the evening sun, a squirrel family lived and played among its low shedding branches; many mornings we were awakened early by Blue Jays chasing those squirrels, the cardinals calling out"whatshere,whatshere". I felt we were living very close to God and his great miracles.

The birds kept very early hours. I loved to lie in bed watching those little feathered creatures flittering from limb to limb, then in the pale violet dusk they would be hunting a limb to sleep on. The nights had a million voices - the "invisible choir", then in the distance a hoot of an owl, one would call then the echo would start dozens of them, many nights sleeping up there so close to nature, I've felt a deep gratitude for the privilege of having these miracles of nature. I've sometimes wished the man in the moon would turn off his light-- but it was beautiful and so delightful, the wind blowing on your face, the morning sunlight sending its golden gleams into our beds to awaken us - can I ever forget it? Nature's secrets and wonders will be with us forever; life can never be dull with so many gifts of nature.

The First World War started soon after we moved to St. Francisville. So many changes came about in the town. Boys and men leaving for training camps and then on across to France. The women were kept busy knitting and making bandages and other services that were needed. The epidemic of flu was a never to be forgotten tragedy; whole families were stricken at the same time. Neighbors helped each other nursing a sick one, bringing in food for each other. We were more fortunate than some of our friends, Gordon and I kept well; Ruffin had a severe case and developed pneumonia. Sister went to help Eliza care for him - so I went to take her place in the Post Office.

Then Gin took sick. Sister came home with the flu - Eleanor came to help me with all these duties and she too went down with flu. Our faithful little cook "Noodie" remained on the job. I don't think I could ever forget this dear black woman. She was so sweet and kind to us all and she came to see us bringing vegetables, chickens and eggs. We were short of doctors during this epidemic. Uncle Feltus had gone into the service and how we did miss him. There were many homes saddened throughout the nation.

Our first and only son - Gordon Jr. - was born in 1921. Gordon was a very proud person, it was amusing to hear him planning the things he wanted to do for "Brother" as he was always called (later in life); the guns he would buy, bicycle, baseball and things he had always liked and still enjoyed; our shady back yard was filled with boys when Brother began to grow up, the many baseball games played, the whole neighborhood gathered to see these games. I think Gin and Brother can look back on these memories; it was the ambition of Gordon to make it happy ones.

Gin entered school at Julius Freyhan; there was no problem getting our children to school as we lived in sight of the school, many times she and Brother dashed out of the door as the school bell rung. There were no interruptions in Gin's school life nor in Brothers, they both finished at the same school they started to.

Every summer our family spent a month at Biloxi, traveling by car; the roads were awful in those days, it often took us several days to make the trip, staying over in country hotels (such awful places some of them were). We carried our lunch basket, full of ticks and red bugs trying to eat in some shady spot along the roadside. Our car was the open type, as most of them were then! It was quite a problem keeping dry when we ran into rain. There would be a mad scramble putting up the curtains, and then we would almost smother trying to get a breath of air when the sun came out, just about the time we got the curtains arranged. Those days weren't called "the good old days"—there wasn't much joy in a motor trip back then.

We did enjoy our vacations in Biloxi, the bathing, fishing and long drives along the beach. Its ever-changing breezes and its song as the waves washed into the sea wall as the tide rose. What a sight those sunsets were and the moon and stars at the night reflecting upon the calm silent waters. I loved to watch the reflecting upon the calm silent waters. I loved to watch the lights from the flounder fishermen's torch as they walked through the night into the shallow water near the shores. These vacations at Biloxi were happy days in our early lives. I can never forget our many trips there. I have never had any great wish to go there again. It would only bring sad memories of happier days.

During Gin's high school days our home was gay with social affairs of her school friends. Our living room and dining room were very large and most suitable for parties. We had a Victrola that furnished the music for their dances. There were Halloween parties, Christmas dances, slumber parties. Summer found the house filled with visitors. The creek or swimming parties by moonlight were great fun. I guess at no time in our life were we any happier than when our children were growing up.

At last Gin was graduating from High School and ready for college. We had to do a lot of planning and figuring though. A depression was upon the nation just when we needed money most. We could use our savings for a while to start Gin to L.S.U. So she entered in September after her graduation. We missed Gin, not having her home, but it was a great satisfaction know-

ing our plans were being carried out and she would have what
neither of us had - a college education.

Gordon, Brother, and I were at home alone now. Brother was
then at an interesting age. Gordon taught him how to handle a
gun, to swim, and today boys become Boy Scouts - their fathers
are deprived of all these pleasures. Gordon loved it; he bought
a tent and the boys camped in our backyard. I remember seeing
a little light burning way into the night, an old kerosene lantern
hung to see by and the boys would be up fighting mosquitoes.
Often, they would take the tent out on the creek and camp for
several days. Gordon and I sometimes drove out to see how
they were fairing, taking a watermelon, cakes and other goodies. There was always an air of excitement and expectancy in
having the boys around. Gordon was living his own boyhood
over again. He loved to tell me of his happy boyhood spent
at Lochimar at Pontiac, Missouri, his grandparents' ancestral
home. How different my childhood had been, losing Mama
when I was only three years old, living here and there, our home
burning, and then off to spend our young lives in a convent.
These things can never be erased from my book of memories.

Gin went to college for four years. There were many loan days
for us during this time. The depression was affecting everyone.
Summer trips were discontinued during these difficult times.
We realized we had to take and accept many obstacles that con-
fronted each and all. Many persons were losing their jobs and
families broken up. We had to face realities in a practical way
with sound common sense.

Eleanor and Joe kept Gin with them in Baton Rouge, so that
she could finish her last year in college. Our old Dodge automo-
bile furnished transportation. Gordon and I sat on our porch
and looked forward to weekends when Gin would return and
the old car could furnish us a ride and change of scenery. Gin

became engaged to Gordon Dippel during her last year at L.S.
U. A boy she had grown up with and had lived across the street
from us, it was a strange and quite a coincidence - after being
away so long and coming home to fall in love with a childhood
playmate. Gin helped Gordon in the drugstore for a while, and
also did substitute teaching at home.

Then, our thoughts were centered on the next big event, Gin's
wedding. We could not afford an elaborate one, neither did
she wish to have one. Many details had to be worked out. We
wanted Gin to have a sweet simple church wedding in Grace
Church where Gordon and I had married and where we had
worshipped so long and many of our dear ones laid to rest
around its sheltering old oaks. Friends and relatives gave parties
for Gordon and Gin, we made shopping trips to New Orleans to
buy the trousseau, and every detail was carried out as planned.

On May 18, 1935, they were married. The choir that both Gin
and Gordon had taken a great interest in sang for their wedding.
They sang "Oh! Perfect Love". The decorations were simple but
with great taste, carrying out Lelia Golson's designs, who was
so kind and thoughtful to undertake this task of love and friendship for me and mine. Many remarked it was one of the sweetest
weddings in Grace Church. We asked a few friends and relatives to a small reception at our home. Aunt Flo Barrow loaned
an old heirloom China punch bowl which had belonged to our
family. Friends sent lovely flowers that made our old house very
attractive. Gin and Gordon went to St. Louis and other points
north to visit and meet her new relatives. When they returned,
Gordon opened the Ford Agency in St. Francisville.

Our lives went on as usual. Again, Gin left us - but only to live
across the street. Brother was still in High School and beginning
to enjoy the girls and parties. He helped Gordon sometimes in
the Drugstore. A year after Gin married my dearest one became

ill. Gordon had not been well for quite a while. He no longer hunted. He seemed worried about the changes taking place, our living habits had changed. These things appeared trivial at the time, but I realized later that he must have felt a foreboding of ill health approaching. So often he mentioned being so tired but would say "Don't worry" – "I'm not sick". So when his last illness and I might add, only illness came on - it was a complete shock. I always pictured growing old together and enjoying our grandchildren in our reclining years.

Gordon died on April 21, 1936, after only a week's illness. I do not like to write of these heart-breaking days. I would prefer not having to write of it at all. There seemed so little of happiness left for me. The future looked dark without Gordon by my side. I had little courage to face the future. I felt I could not live to face this strange frightening and new world before me. I could not flee from the circumstances and conditions confronting me. I had to face them. A deep faith in God was necessary. I needed help in facing myself. All my life I had leaned on someone. I had no shoulder to lean on now. Only God, to whom I prayed fervently to give me strength and courage. I found this courage and strength in my church. I found peace and help in humble prayers.

I knew nothing of operating a drugstore. Gordon had always advised me to sell the store in his absence should he go first. I wondered how I would make a living for Brother and myself. I decided to keep the store for a while. Mr. David Walker of New Orleans was sent to me by I. Lyons & Co. to manage the store. He was a fine old gentleman of the "Old School" type. Mr. Walker boarded with me and was a good friend, we all were very fond of him, but he became ill and then my real headaches began. There was a series of changes, whiskey heads, drifters; the worst kinds came to work for me. Brother was in school but helped at times in the store. Poor little fellow was deprived of a father and the advantages of an education; he would have to go to work when high school was finished. Life was pretty hard for both of us. I look back now and wonder how I lived through these hectic days. Sister and all the others were so fortunate in having my dear children and good friends.

Gin and Gordon decided to come and live with me. I rented most of my house and gave meals to those who stayed with me. Most of those in my house were young people. I was kept very busy and the house was quite gay with their jokes and laughter. It was the best thing that could have happened to me. I seldom had a quiet moment. In March 1937 Gin and Gordon's first baby came. Born in the old brick house we all loved. A baby brings many changes into a home. Life became most interesting for us all. "Little Ginger" was an adorable chubby baby, strong and healthy. The pet of the house. We all enjoyed her.

Gordon, Jr. was now about to graduate. What was I to do with him? I was proud to know he would finish high school. He found small jobs here and there during the first year after graduation. Then he decided to enlist in the Marines. I can never forget the early morning he left to go to New Orleans, to enlist. I listened to the old train going down the track by our house. How many times that same train had whistled and awakened him as a baby; now it was carrying him far away. Would I ever see him again? I shed tears that morning in my pillow. No one knew how my heart ached.

When the train came in that night, I listened and wondered, would he be back?? Well in a few minutes there stood Brother with his little handbag waiting to be let in the door. I had not realized what his leaving meant. It was God sending him back home to me. In a short time after this World War II was declared, he probably would have been among the missing on Okinawa along with many other poor boys who lost their lives on this poorly fortified island. I have never ceased to be thankful that he returned home (I guess those bad teeth saved his life) for that was why he was rejected. After working about at various small jobs Brother went to L.S.U. to take a course in the Air Corp. The Government had started a training project for the boys unable to pay tuition. This project did not prove successful and it was discontinued in a short time after Brother entered. The C.O. advised him to go to Barksdale Field and enlist in the regular Air Force. He was given a nice letter of recommendation and was accepted, thus began his military career. As I write this he is now going into his twelfth year in the service.

In 1941 Brother married his boyhood sweetheart, Kathryne Walker. They were married Christmas Day. War had been declared with Japan. All the boys were marrying before being shipped to foreign countries. Kate was in training in Baton Rouge. He and she went to Albuquerque, New Mexico to live. I won't write too much about those distressing war days. It was pretty much of a repetition of World War 1. Homes were broken up, wives were following their husbands around the army camps, leaving their babies with the Grandparents. I'll never forget the food rationing days. There was a mad scramble to get food and wearing apparel. Women almost fought over nylon hose, a piece of ham, canned fruit and bed linens. I was fortunate in many ways, but what a headache I had trying to feed my family and boarders with food restrictions. I still dislike Spam and cheap sausage. They were our main meat dishes for supper. Ham was a great luxury.

A new baby came to Gordon and Gin in 1940. Hazel was born. She was a dear little girl, strong and healthy, but not as fat as Ginger. I was trying to run the drugstore, keeping the boarders fed. We were all very busy, everybody worked. I had a faithful servant - Ellen – who cooked dinner and helped clean the house. I don't think I will ever forget Ellen's faithfulness to me.

She seldom missed a day and walked two or more miles back and forth to her work. Kathryne and Brother's first baby was coming. I was unable to go to them so Eleanor went out to Albuquerque to help with the new baby. Mary Kay was born May 1943, and eighteen months later, another baby came to them. This time I arranged to go out and Virginia and Ellen took over the house for me.

My trip to Albuquerque was a series of delays, missing connections, trying to get reservations, waiting in dirty railroad stations, with a war going on, these things were expected, one traveled at your own risk. It took me three days to make the trip. I don't think I could ever forget the trip. I liked Albuquerque very much. It's lovely bright sunny days, it was December. I spent my first Christmas away from home. It was a very happy one. I cooked a big turkey that took us quite a while to eat. There was a pretty little Christmas tree for Mary Kay and Bonnie, the new baby. Brother and Kathryne's little house was the adobe type - very comfortable and attractive. We could see snowcapped mountains from the living room windows. It made a beautiful picture and such a wonder to me. I enjoyed every minute of my visit. I found it interesting to see the many things made by Indians, jewelry, beads and rugs. It amused me to meet them about the streets, shopping in the stores, jabbering and spending money freely and driving fine cars. Their clothes were real Indian style. Gay head bands, blankets over their shoulders, long black hair braided down their backs.

I truly got a big kick out of these unusual sights. I hated to leave when my visit came to an end. I knew I must go home. Kathryne and I wept when the time came to go. She had to care for two babies now and so far away from all of us. When the taxi came to get me, that cold winter night, she was clinging to me and Brother stood and watched us both cry. My heart was heavy that night. I had no idea when I would see them again.

Orders were corning every day for men to go overseas. Brother knew his time wasn't far off.

An indelible imprint on my memory was the remembrance of Brother standing there as the train pulled out from the station; it was bitter cold; all I could see was his tall figure waving to me - his face buried in his high collared overcoat. How I dreaded that trip. I witnessed many sad scenes, wives, mothers, and sweethearts bidding good-by all along the route to their loved ones, boys leaving their families behind. Some of them never forget some of these scenes. I had a lump in my throat as I rode through the night.

I was happy to get back home and surprised to learn on my arrival that little "Miss Hazel" had set the house on fire while I was gone. It was quite an exciting time. Gin was all alone and it gave her an awful fright, fortunately help came quickly and little damage was done. Some furniture was damaged and one room had to be redone. I was not told of all this until I returned.

Hazel was always very venturesome, so often getting into many predicaments. One time she fell into Sister's goldfish pond. This could have been a tragedy.

Brother's orders to go overseas came. Their sweet little home had to be broken up. This was only a few weeks after I left them. They came to stay a short while with me. Kathryne and the babies went to visit her folks awhile then came to remain with me while Brother was gone.

My old house was stretched to its capacity. Gins third child was on its way. Danny was born in June 1945. We were all so happy over the birth of a boy.

Life was very interesting at my house. I never knew where I would sleep. Baby beds in every room. Ice box filled with babies' formulas. I can still recollect Gin and Kathryne fixing milk bottles. There was scarcely any room in the ice box for drinking water. I remember one time I was babysitter for one of the babies and got the formulas mixed up. I gave Bonnie's bottle to Danny and vice versa. However, it did no harm. We all had a good laugh over it. I had several boarders, mostly girls; there was an air of gaiety over the house with so much youth about.

The war was raging, both in Japan and Europe. Everyone was too worried to fret over inconveniences or short wages, how fortunate we felt here in America, no bombs being dropped on our homes, our children safe and healthy. God Bless America, how thankful we all should be for our many blessings. An air of suspense hung over each and every home; one never knew when a fatal message would come.

Brother sailed from San Francisco for the Philippine Islands. He did not encounter many dangers except while making the trip there was always danger from submarines. He had some very interesting experiences in Manila. He brought back lovely gifts for us all. After a few months over there the war was ended and he was returning home, and was back with us all. The joy of having peace restored was untold. It took months to read-just to normal living and food rationing would be discontinued. Prices, however, have returned to normal as I write this (nine years later) 1953 - everyone is enjoying an era of prosperity.

I was compelled to sell the drugstore at the close of the war. I could not find anyone to run the store for the wages that I could afford to pay. So for the first time in many years the little drugstore changed hands. Ruffin had worked in that store for many years when he was first married – later – Jon (Gordon's brother) had worked there, then Gordon had been there for over seventeen years. It just seemed part of me, seeing Gordon there standing behind the prescription counter or bending over his desk figuring on his books. He loved the store; it was a second home to him, a gathering place of the gents of the time.

The school children gathered at noon for candy and drinks. There was always an air of friendliness about the place. So I felt a great sentiment towards what had been very dear to him. It was a great relief to me without this responsibility. I had tried to give the public the same service that Gordon had. My true friends stood by me in spite of the many obstacles I encountered. I missed the public, but I went to work for a while as a clerk helping Mr. Ferguson, when I sold it. When his son returned from the war, he took over the place as assistant to his father, so I was no longer needed. I then had to find other means of earning my living. I had my home and the rooms and boarders which took care of some of my needs.

Brother re-enlisted in the Air Corp. after enjoying a three-month furlough visiting us and other relatives. He and Kathryne and the children went to live in Dayline, Louisiana near Barksdale Field where his services were. The house was rather empty without them. Gin and her family were still with me. Another baby was expected. Gordon Edward was born in April:-1946. So then there were four children in the Dippel family. Gordon and Virginia decided to buy a place just out of town. I can still remember each night Gordon and Gin drawing plans and figuring over the new home cost. Materials were hard to get as prices were still very high, but building a home was now a necessity for them as their family had outgrown my small quarters downstairs. At last the plans were completed, the house was completed.

I was faced with being left alone. The first time in my entire life. It was truly a new experience. I realized what it was to face life alone. I rejoiced with Gordon and Gin in having a home of their own. I knew I would miss them. It would be difficult to adjust myself to my new way of living. There is an unfolding of life within each of us that is known only to ourselves. It would be wonderful if we could have a sundial at the door of our heart so that it could only record the sunny hours. I guess I felt pretty sorry for myself at that time. No one knew the little heart pangs that tugged at my heart. I was alone. What next? Just at that critical time - Brother and Kathryne came for a visit for two weeks. I was so very happy!! I needed something to pep me up and it came at the right time. After pondering over the future, I decided to set up housekeeping unit in the rooms that wore loft empty downstairs. A nice young couple came to live with me and I was not alone.

Gin's home was very large and comfortable. Lots of space where the children could have pets, chickens, a garden - all the things that make a home. The children were very happy and Gin and Gordon were kept very busy and interested. I guess I too enjoyed the relaxation, quiet and peace that I perhaps needed and had not had for many years.

I was offered a position as a hostess at "Afton Villa", soon after I had settled down to this quite way of life. I was so pleased that I could find some interesting work suitable for one of my years. I would meet people and enjoy my new environment amid beautiful surroundings that had once been the home of my great grandfather David Barrow now owned by Mr. and Mrs. Wallace Percy. The history of this old house fades back into a glamorous past. A French chateau, built to please a winsome young bride. As I write this story of my life, I am sitting by a window overlooking a terraced garden at Afton Villa azaleas, camellias, sweet olives, magnolia all make up this peaceful view. A large oak tree with its moss hung branches make a soft shade where I sit. A peacock with its colorful feather spread into a graceful fan, struts just outside my window. It calls to its mate, making a loud worrisome sound. When I return home in the evenings after my days' work is done, I find peace and contentment in the four walls of my home. Its charm and dignified atmosphere is what I love.

I have a very dear friend who shares my home and brings gaiety and youth into my life. Bonnie Davy and I have been living together a number of years, both of us being widows we have quite a lot in common. There is quite a difference in our age, yet we share a congenial friendship. My family feels very gratified that I am not alone and have a good true friend to share my home. Bonnie is good company, full of humor, always a joke or some bit of news to tell me, these are the things I need. My happiness and companionship with Bonnie is a consolation and comfort against the older years when people grow lonely. What have I contributed to make her life more interesting? I do not know; however, I have tried to make my home her home too, the flowers and garden, I practically have turned over to Bonnie her friends and relatives I've tried to make welcome. Sister and Tibbie who live just at the back of my house have grown old together. I can see the light go out each night, a great comfort to know they are near me.

Ruffin, like me, lives alone except for some boys who rent from him. Eliza died in 1951. He has a comfortable home in Baton Rouge and though infirm and mostly confined to his home he does not want for the comforts of life. Eleanor and Joe live near and keep in close touch with Ruffin. They have a nice home and enjoy life as the years go by. Their home is a favorite stopping place for us all on our trips to Baton Rouge. Brother and Kathryne are now living in Baton Rouge and Brother is still working in uniform for "Uncle Sam".

I see Virginia and her family every day. Our younger brother Floyd lives in Jackson, Tennessee. He is married and has one dear little girl, Jane. They come every summer to visit me. Papa and Edna are both buried at Highland, where Mama and Bird are buried. I don't want to dwell too much now on the past (that's a real sign of old age). Providence has given me a good life. I've been loved and cherished through the years by my dear sisters and brothers and my children and grandchildren. I've traveled to a Golden Age.

Life is so complex but there is always a shaft of light piercing through it. The Golden Years can be happy and profitable years. I trust mine will continue to be as I move into retirement with unfaltering faith of an unknown future. Youth once it has fled cannot be recaptured except in memory. I thank God that I have kept my memories and that I have now come to the "GOLDEN AGE".

NOTE: My dear "Mudie" passed away on Jan. 15, 1955, at the age of 66 years. She died of a heart attack which came very suddenly. It was just as she would have wished, quick and final. She was only sick a week, she had her first attack on January 9th while attending a dinner at Jackson Hall after church services. We took her to the Baton Rouge General Hospital. The next day she seemed to be getting along so fine. She had no warning of death. She had the last attack on Saturday, Jan. 15th and went into unconsciousness immediately. She suffered no pain of struggle and died quietly. My Aunt Jo was with her and also Brother, but she did not know anything. I am so thankful she wrote this story of her life; in a way it is the story of my life too.

I feel now that my dear Mother and Daddy are united and happy together again.

Gin —

Jan. 18, 1955

THE END

ABOUT THE AUTHOR

ALSTON M. (MAC) BARROW worked in the financial industry for fifty years including institutional bond sales and stocks on Wall Street, insurance, investment advisory, and newsletter publishing.

He is now retired and lives in Tampa, Florida with his wife along with his children and their families, fourteen in all. He has cattle ranch property in Florida and Colorado where he and members of his family enjoy horseback riding, hunting, and fishing. He also loves cooking, vegetable and herb gardening, and travel.

9 798991 516723